Stop Faking It!

Finally Understanding Science So You Can Teach It

CHEMISTRY BASICS

CHEMISTRY BASICS

Arlington, Virginia

Claire Reinburg, Director
Judy Cusick, Senior Editor
Andrew Cocke, Associate Editor
Betty Smith, Associate Editor
Robin Allan, Book Acquisitions Coordinator

ART AND DESIGN Will Thomas, Jr., Director
Brian Diskin, Illustrator
PRINTING AND PRODUCTION Catherine Lorrain, Director
Nguyet Tran, Assistant Production Manager
Jack Parker, Electronic Prepress Technician
*sci*LINKS Tyson Brown, Director
David Anderson, Database and Web Development Coordinator

NATIONAL SCIENCE TEACHERS ASSOCIATION
Gerald F. Wheeler, Executive Director
David Beacom, Publisher

09 08 07 4 3 2 1

Library of Congress Cataloging-in-Publication Data
Robertson, William C.
Chemistry basics / by William C. Robertson, in consultation with Michael S. Kralik ; illustrations by Brian Diskin.
p. cm. -- (Stop faking it! : finally understanding science so you can teach it)
Includes index.
ISBN 978-0-87355-239-4
1. Chemistry--Popular works. I. Title.
QD37.R53 2007
540.71--dc22
2006037136

NSTA is committed to publishing quality materials that promote the best in inquiry-based science education. However, conditions of actual use may vary and the safety procedures and practices described in this book are intended to serve only as a guide. Additional precautionary measures may be required. NSTA and the author(s) do not warrant or represent that the procedure and practices in this book meet any safety code or standard or federal, state, or local regulations. NSTA and the author(s) disclaim any liability for personal injury or damage to property arising out of or relating to the use of this book including any recommendations, instructions, or materials contained therein.

Contents

Preface

The book you have in your hands is the eighth in the *Stop Faking It!* series. The previous seven books have been well received, mainly because they stick to the principles outlined below. All across the country, teachers, parents, and home-schoolers are faced with helping other people understand subjects—science and math—that they don't really understand themselves. When I speak of understanding, I'm not talking about what rules and formulas to apply when, but rather knowing the meaning behind all the rules, formulas, and procedures. I know that it is possible for science and math to make sense at a deep level—deep enough that you can teach it to others with confidence and comfort.

Why do science and math have such a bad reputation as being so difficult? What makes them so difficult to understand? Well, my contention is that science and math are *not* difficult to understand. It's just that from kindergarten through graduate school, we present the material *way* too fast and at too abstract a level. To truly understand science and math, you need *time* to wrap your mind around the concepts. However, very little science and math instruction allows that necessary time. Unless you have the knack for understanding abstract ideas in a quick presentation, you can quickly fall behind as the material flies over your head. Unfortunately, the solution many people use to keep from falling behind is to *memorize* the material. Memorizing your way through the material is a sure-fire way to feel uncomfortable when it comes time to teach the material to others. You have a difficult time answering questions that aren't stated explicitly in the textbook, you feel inadequate, and let's face it—it just isn't any fun!

So, how do you go about *understanding* science and math? You could pick up a high school or college science textbook and do your best to plow through the ideas, but that can get discouraging quickly. You could plunk down a few bucks and take an introductory college course, but you might be smack in the middle of a too-much-material-too-fast situation. Chances are, also, that the undergraduate credit you would earn wouldn't do the tiniest thing to help you on your teaching pay scale. Elementary and middle school textbooks generally include brief explanations of the concepts, but the emphasis is definitely on the word *brief,* and the number of errors in those explanations is higher than it should be. Finally, you can pick up one or fifty "resource" books that contain many cool classroom activities but also include too brief, sometimes incorrect, and vocabulary-laden explanations.

Given the above situation, I decided to write a series of books that would solve many of these problems. Each book covers a relatively small area of science, and the presentation is slow, coherent, and hopefully funny in at least a few places. Typically, I spend a chapter or two covering material that might take up a paragraph or a page in a standard science book. My hope is that people will take it slow and digest, rather than memorize, the material.

This eighth book in the series is an introduction to chemistry basics, and the order and fashion in which I present the concepts is most likely different from what you have seen in other chemistry books. I spend quite a bit of time on the structure of the atom and why we believe that atoms look a certain way. I believe that you can't really understand chemical reactions without a thorough understanding of atomic structure. In addressing the structure of the atom, I include results from quantum mechanics. This might seem odd in a "basic" chemistry book. These results, however, are crucial to our understanding of the atom. Also, there are a number of simple ideas one can draw from quantum mechanics without getting into all sorts of messy math. So, no differential equations in this book! I am developing a follow-up chemistry book that goes into more depth in many areas of chemistry, so just for kicks I might toss in some of that messy math at that time.

There is an established method for helping people learn concepts, and that method is known as the Learning Cycle. Basically, it consists of having someone do a hands-on activity or two, or even just think about various questions or situations, followed by explanations based on those activities. By connecting new concepts to existing ideas, activities, or experiences, people tend to develop understanding (unlike with memorization). Each chapter in this book, then, is broken up into two kinds of sections. One kind of section is titled, "Things to do before you read the science stuff," and the other is titled, "The science stuff." If you actually do the things I ask prior to reading the science, I guarantee you'll have a more satisfying experience and a better chance of grasping the material. Keep in mind that the activities in this book are designed to help with *your* understanding of chemistry. Although you might decide to use some of the activities with students, that's not their main purpose.

It is important that you realize the book you have in your hands is *not* a textbook. It is, however, designed to help you "get " science at a level you never thought possible, and also to bring you to the point where tackling more traditional science resources won't be a terrifying, lump-in-your-throat, I-don't-think-I'll-survive experience.

One more thing. I often get comments from teachers something like the following: "But I don't teach quantum mechanics (or binary electric circuits, or calculus, or whatever) in middle or elementary school. Why should I learn the basics of these subjects?" Once again, it's about comfort level. If you know more than you are required to teach, you have a better idea of how to approach classroom discussions and how to answer certain questions. In other words, you are in charge rather than the classroom materials being in charge.

Dedication

I dedicate this book to Norris Harms, who convinced me to quit working for other people and strike out on my own. It hasn't been easy, but it's been worth it. I also dedicate this book to Rob Warren, who taught me how to write directly and clearly and drop the scholarly sounding tone and flowery language.

About the Author

As the author of NSTA Press's *Stop Faking It!* series, Bill Robertson believes science can be both accessible and fun—if it's presented so that people can readily understand it. Robertson is a science education writer, reviews and edits science materials, and frequently conducts inservice teacher workshops as well as seminars at NSTA conferences. He has also taught physics and developed K–12 science curricula, teacher materials, and award-winning science kits. He earned a master's degree in physics from the University of Illinois and a PhD in science education from the University of Colorado.

About the Consultant

Michael Kralik received his PhD in chemistry from the University of Utah with post doctorate studies in chemistry, pharmacology, and toxicology. He has been faculty at the university and has conducted many faculty and staff development seminars. Kralik has established product development and manufacturing operations domestically and internationally, and has directed the development of hundreds of products for major corporations in chemical, medical, pharmaceutical, and electronics industries. He has developed K–12 science curricula, teacher inservice workshops, and many award-winning educational toys, games, and science kits.

About the Illustrator

The recently-out-of-debt, soon-to-be-famous, humorous illustrator Brian Diskin grew up outside of Chicago. He graduated from Northern Illinois University with a degree in commercial illustration, after which he taught himself cartooning. His art has appeared in many books, including *The Beerbellie Diet* and *How a*

Real Locomotive Works. You can also find his art in newspapers, on greeting cards, on T-shirts, and on refrigerators. At any given time he can be found teaching watercolors and cartooning, and hopefully working on his ever-expanding series of *Stop Faking It!* books. You can view his work at *www.briandiskin.com*.

About This Book

What you have before you is a book intended to help you understand many of the concepts basic to the study of chemistry. This is not a chemistry textbook, and it takes an approach that differs from that of most of the books on chemistry you'll find. For starters, this is a view of chemistry from someone (me) who has primarily a physics background. This means that I will take a "ground up" approach to the subject, beginning with what atoms look like and why we believe that atoms look that way. This does not follow the historical development of modern chemistry, but it's the approach that makes the most sense to me. I also will steer away from memorization whenever possible. I see no sense in memorizing the properties of various materials based on their position in the Periodic Table when you can rather get a clear picture of why materials do what they do based on an understanding of the mechanisms that underlie the patterns in the Periodic Table.

There will be a second book on chemistry in the *Stop Faking It!* series, so don't fret that there are a number of traditional concepts missing from this book. People might argue about what concepts are "basic" and therefore should be in this first book, but I just have to chalk that up to a difference of opinion.

I do make a few assumptions about what you already know before you pick up this book. For example, I assume you know that there are three states of matter—solids, liquids, and gases—and that you have a rough idea of the properties of each state of matter. I also assume you know that when you combine atoms together, the result is called a molecule. Other than that, I pretty much start from scratch.

All that said, here are the major ideas I hope you can grasp by the time you finish this book:

- What an element is, both historically and presently
- Our current view of what is inside an atom, and what electrons are or aren't doing inside an atom
- How energy is involved when physical systems, including atoms, are by themselves and how energy is involved when those systems get together

- The different ways, and different reasons, atoms bond with one another
- What we mean by a chemical equation, how to balance a chemical equation, and how we can get valuable information from a chemical equation. This includes why in the world someone would want to balance a chemical equation, other than to do another textbook chemistry problem.
- What we mean by organic chemistry, why it is a special branch of chemistry, and how it affects our lives

Warning. This book is intended as a resource book for teachers and parents, and not necessarily a book of activities designed for the classroom. The use of certain chemicals in the classroom is highly regulated in many areas, and I do not recommend doing the activities in this book with students without checking on the regulations in your particular area.

Safety Note

Though the activities in this volume don't require anything more volatile than household vinegar, safety should always be in the forefront of the mind of every teacher. (This is not intended as a book of classroom activities, by the way. Rather, the activities are designed to enhance your understanding of the subject before you get into the classroom.) Your individual school, or possibly the county school system of which your school is a part, likely has rules and procedures for classroom and laboratory safety.

You can also find specific guidelines for the safe storage, use, and disposal of thousands of types of chemical products in the Material Safety Data Sheets (MSDS). Start with *http://www.ilpi.com/msds/#Internet*. This site links to dozens of free searchable databases, including those of top American and European universities.

NSTA has also published several award-winning titles covering the safety theme at all school levels. For the elementary level, there's *Exploring Safely: A Guide for Elementary Teachers* and the *Safety in the Elementary Science Classroom* flipchart. Middle school-level offerings are *Inquiring Safely: A Guide for Middle School Teachers* and the *Safety in the Middle School Science Classroom* flipchart. Finally, there's *Investigating Safely: A Guide for High School Teachers*.

How can you avoid searching hundreds of science websites to locate the best sources of information on a given topic? SciLinks, created and maintained by the National Science Teachers Association (NSTA), has the answer.

In a SciLinked text, such as this one, you'll find a logo and keyword near a concept, a URL (*www.scilinks.org*), and a keyword code. Simply go to the SciLinks website, type in the code, and receive an annotated listing of as many as 15 web pages—all of which have gone through an extensive review process conducted by a team of science educators. SciLinks is your best source of pertinent, trustworthy internet links on subjects from astronomy to zoology.

Need more information? Take a tour—*www.scilinks.org/tour*

Simple Models

As I said in "About This Book," I'm not going to take the usual approach to the subject of chemistry. Because virtually all explanations of chemical reactions are based on our current model of atoms and molecules, I figure the first thing to do is help you understand why we believe that atoms and molecules look and act the way they do. That's not a trivial issue, because despite the impression you might have gotten from textbooks, no one has ever seen an atom in the sense that you can see this page in front of you. What we have are observations and experiments that lead us to formulate models of atoms.

"Fire. Definitely 'Fire'..."

For starters, I'm going to have you experience a few chemical reactions and then present a model the early Greeks used to explain observations. That'll get us in the mood for making up models to explain the world around us.

Things to do before you read the science stuff

In this section, you'll need a few chemicals that you can get at the store—baking soda, vinegar, and Epsom salts. You'll also need a couple of metal bottle caps with the plastic inner liner removed, a ruler, a pencil, tongs, wooden matches, a candle, and a lighter. Ready to play? Good. Now, before you do anything, prepare to observe carefully. Pretend that your life depends on how detailed your observations are. Okay, that's too melodramatic, but observe carefully anyway, okay?

Pour a small amount of baking soda in a clear glass or plastic cup. With the cup sitting on a table, add maybe 40 milliliters (an ounce or two) of vinegar. Without disturbing the cup, describe what happens. What do you see and hear? Now light a candle and place it next to the cup. Carefully "pour" the contents of the cup onto the candle. "Pour" is in quotes because I don't want you to pour out any of the liquid in the cup. Just use a pouring motion until something happens to the candle. See Figure 1.1.

Figure 1.1

For the next couple of activities, a double pan balance is nice to have. If you don't have one, make one as follows. Get two bottle caps and use masking tape to secure them to opposite ends of a ruler. Then place the ruler on top of a pencil, with the pencil in the middle of the ruler, and adjust the positions of the bottle caps until this whole thing balances. See Figure 1.2.

Figure 1.2

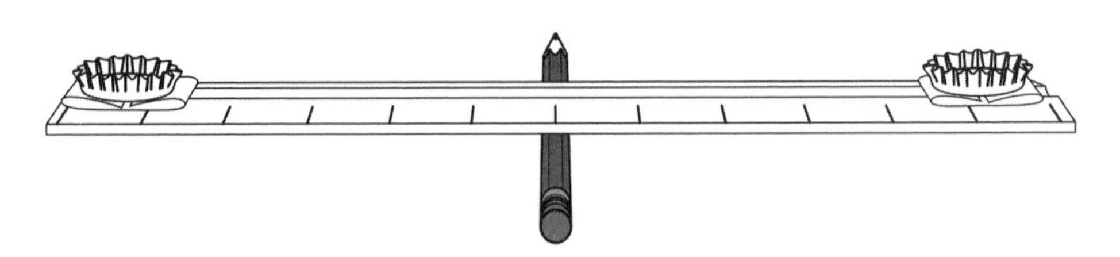

Figure 1.3

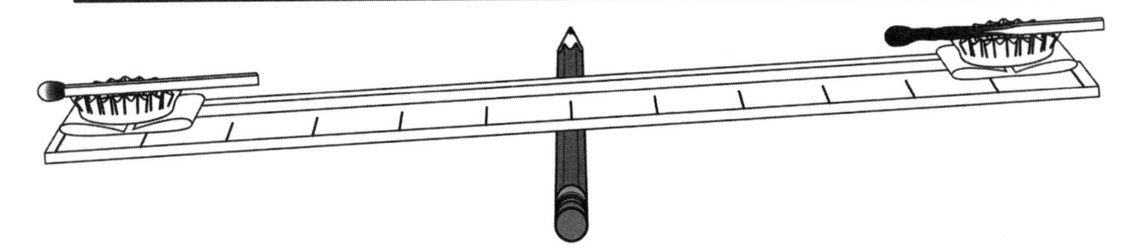

Place one wooden match on top of each bottle cap (or in each pan if you have an actual double pan balance) and check that your apparatus still balances. Strike a third match and light one of the matches sitting on the ruler on fire. Make careful observations of the burning process. Which way is the flame directed? Does anything besides the flame seem to "leave" the match as it burns? What about after the one match completely burns? Does your apparatus stay in balance? If not, which way does it tip? (See Figure 1.3.)

Get another match and a small glass. Light the match and hold the glass, inverted, above the match, as shown in Figure 1.4.

Figure 1.4

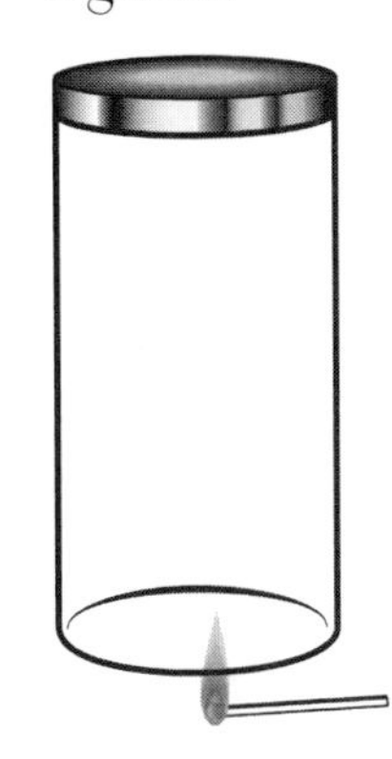

Before the match burns your fingers, blow it out. Then take a look at the inside of the glass. Rub your finger on the inside of the glass. Notice anything?

Clean out your bottle caps and get your balance in balance again. Pour a small pile of Epsom salts into each bottle cap, adjusting the amounts until your apparatus is in balance. If you're using a regular pan balance, you still need to use bottle caps for this part. Simply place the Epsom salts into bottle caps and then place the bottle caps on the balance pans. No masking tape necessary unless you're using a ruler for a balance. Check out Figure 1.5.

Figure 1.5

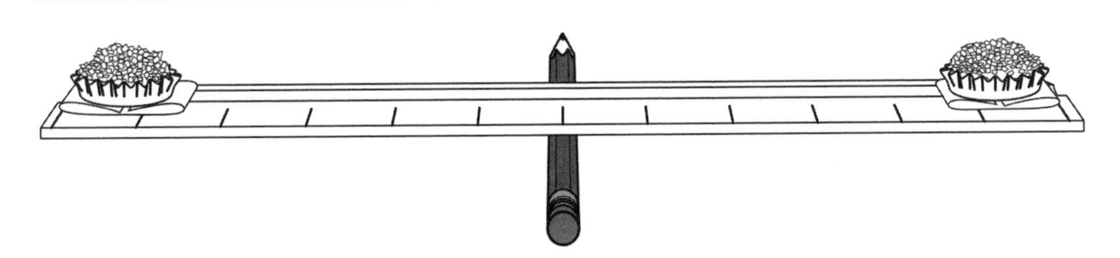

Next remove one of the bottle caps containing Epsom salts. Using tongs so you don't burn your little fingers, hold the bottle cap over a flame (either a candle or a lighter) for a few minutes. Notice any change in the appearance of the Epsom salts. Notice anything else while you're heating things up?

Figure 1.6

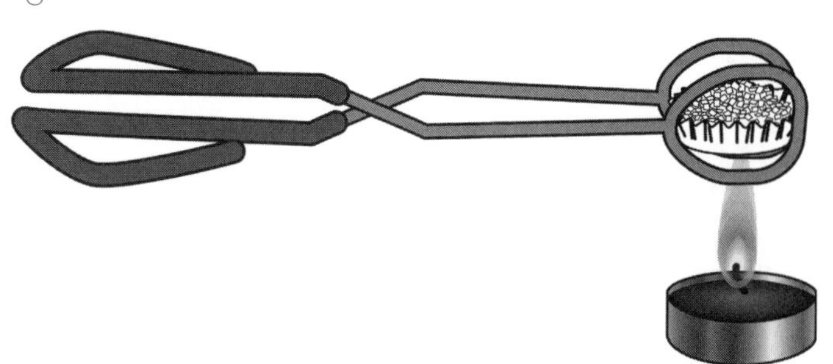

After heating for a few minutes, let the bottle cap cool down a bit and then replace it on either your ruler or your double pan balance. Which way does the balance tip? What does that mean in terms of the relative weights of the two sides?

The science stuff

Before getting into explanations, see if your results agree with the following. As you do this, you no doubt will realize that, throughout this book, I'm going to be discussing what most likely happened in the "things to do" sections. Therefore, you can get through the book without ever doing a single activity other than reading. That, however, is not the best way to learn concepts. Actually experiencing things will give you a solid basis for understanding concepts, a better basis than you can get just by reading. Okay, end of lecture. Here's what you probably observed.

- Vinegar has a distinctive smell. Baking soda doesn't. When you combine the two, lots of fizzing and bubbles result. If you put your hand or cheek just above all the fizzing, you'll notice that *something* seems to be moving upward, because liquid keeps spitting at you.
- "Pouring" the baking soda and vinegar mix over the candle results in the flame being snuffed out. You can't see anything actually pouring over the candle.
- A burned match weighs less than an unburned match, as evidenced by the fact that your balance, be it ruler or actual pan balance, went down on the side of the unburned match. The flame of the match rises upward.
- When you hold a burning match underneath a small glass, a mist of water forms on the inside of the glass. You can usually see this, but you can always feel it when you rub your finger on the inside of the glass.
- When you heat Epsom salts, they sizzle quite a bit. You can definitely see some sort of liquid form as you heat the salts, and they change appearance from crystalline to something that's more of a powder. You should have found that Epsom salts lose weight when you heat them.

Okay, time to explain your observations. As promised in the introduction to this chapter, we'll use a model developed by a Greek philosopher, in particular one named Empedocles. His model was that all things in the universe consist of four elements: earth, water, air, and fire. These elements are characterized by the following qualities: earth is cold and dry, water is cold and moist, air is hot and moist, and fire is hot and dry. Empedocles went even further, assigning qualities such as love and hate to these elements, but we won't go into that other than to say that love and hate are emotions often associated with chemistry and other sciences.

According to Empedocles, each object or substance is some combination of the four elements. The elements themselves never exist in a pure form, although they aspire to take their proper place in the universe. The proper place for earth is, well, at the Earth. Because water floats on earth, its proper place is above the Earth. Air rises higher than water, so air is above the water. Fire rises highest of all (flames go upward, right?), so it's on top. Figure 1.7 shows how Empedocles might have viewed things.

Figure 1.7

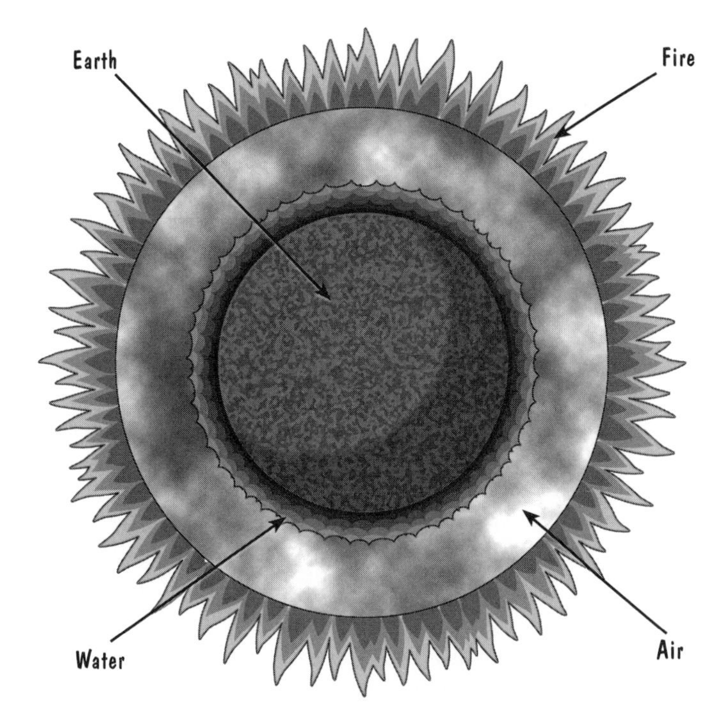

You can explain lots of occurrences by assuming that things are composed of these elements. I'll attempt to do that for the activities you did in the previous section. No guarantees that my explanations are what Greek philosophers would have provided, but they're dead and can't do anything about it.[1]

[1] Irreverence aside, I want to be clear that I am not a qualified scientific historian. The explanations I'm going to provide are most assuredly not the ones Empedocles or other Greek philosophers would have provided (for one thing, mine are too simple), and of course there were many theories that competed with the one of earth, fire, air, and water. My only purpose here is to show that seemingly primitive models can go a long way toward explaining observations, even in the hands of a rank amateur like me.

So here goes my explanation of the activities assuming the Greek notion of there being four elements—earth, water, air, and fire. Let's start with baking soda and vinegar. The vinegar itself obviously has lots of water in it, but it must also contain air. If it didn't contain air, then the smell from the vinegar wouldn't be able to rise up to your nose. Keep in mind we're talking about air as one of the four basic elements. If there isn't any air or any fire, then nothing rises up. When you mix the two, you get more evidence of "trapped air." Those bubbles rise upward and release air. Most likely, the baking soda itself also contains the element air. The fact that you can "pour" an invisible substance out of the cup and extinguish a candle is a bit more difficult to explain, but not too hard. Clearly whatever is right above the mixture of baking soda and vinegar has air (because it's invisible) and either a bit of earth or water, or both. We know there is either earth or water present, because otherwise this stuff would move upward just like all other things that have lots of air. There's enough water and/or air, in fact, that this substance falls downward and extinguishes the candle. When the candle is extinguished, clearly that invisible falling substance has absorbed the fire in the candle.

Let's move on to a burning match. An unlit match contains earth (obviously) and also contains fire. You know it contains fire because simply striking the match releases the fire. The evidence that a match also contains water is in the formation of a fine mist of water on the inside of a glass held upside down over a burning match. Okay, why should a burned match weigh less than an unburned match? Well, you have released fire and water from the match, so there's less "stuff" in the match. Of course, it's not quite that simple. Because fire tends to move upward, you might expect the release of the "upward tendency" of the fire would actually cause the match to weigh more. However, we can get around that by noticing that water would contribute to the match's weight, so as long as you release more water than fire, the match will weigh less.

I'm going to leave the explanation of the Epsom salts to you. I'm sure you can do as well as I can simply by considering the release of some of the four elements that make up the Epsom salts. I *can* add one thing, which is that the salts go from a bright, crystalline substance to a powdery substance. The early Greeks associated light with fire, so as the crystals lose some of their luster, they are obviously losing fire.

SCILINKS. THE WORLD'S A CLICK AWAY

Topic: Origin of Elements

Go to: *www.scilinks.org*

Code: CB001

Time to move on to a different kind of simple model to explain observations. Another Greek philosopher by the name of Democritus was among the first to come up with the theory of atoms—small, indivisible things that make up all matter. In Democritus's view, and in the view of many scientists who also subscribed to the atomic theory into the eighteenth century, there are many kinds of atoms in the world. **Elements** are substances that are composed of only one kind of atom, and observations can be explained by the exchange of atoms among substances. Contrary to the theory of Empedocles, who thought that "pure" elements did not exist in the real world, Democritus tells us it is now possible to have *pure* elements that are composed of only one kind of atom. Like the theory of Empedocles, however, explanations amount to figuring out how much of each element is contained in a substance, how much might be added, and how much might be subtracted. Let's take the heating of Epsom salts as an example. The Epsom salts lose weight when you heat them. Obviously that means that you have fewer atoms after heating than before heating. The liquid that forms during heating obviously indicates the release of some kind of water-like atom that was trapped in the salts prior to heating.

Although it might seem that the early atomic model of Democritus, being closer to what we believe today, is better than earth, air, fire, and water, it's actually not as good. At least with Empedocles' model you could figure out the properties things might have based on how much of each of the four elements they contained. Small, indivisible atoms don't even offer that much, especially if the known list of elements is rather small. Of course, the point of this chapter is not to convince you that early models were good ones. My main point is to illustrate, with a few examples, how people used models to try and explain real-world observations. We do the same thing today, even though our models are a bit more complicated.

More things to do before you read more science stuff

In order to introduce more complex models, I'm going to have you do a few more things. To start with, grab a balloon, blow it up, and tie it off. Then rub it on your hair and move it away from your hair. Unless you have a lot of gel or hairspray in your hair, or if you don't *have* much hair, your hair should move toward the balloon as you pull it away.

Now get an empty aluminum can and place it on a hard surface. Rub that balloon on your hair again and then slowly bring it near the can. What happens?

More science stuff

If you already know a bit about static electricity, you might have an explanation in hand for what you just observed. If you rely on Empedocles or Democritus for your explanation, though, you're going to have difficulty. Nothing in those models explains how one object can exert a force on another object without even touching it. You'd have to chalk it up to the influence of the gods, which of course would have worked just fine for the Greeks.

The kinds of interactions you observed have been known for a long time (since the eighteenth century), and they led to alterations of the model of the

Topic: Atoms and Elements
Go to: *www.scilinks.org*
Code: CB002

atom. Scientists proposed that there were two kinds of charges in the world—positive and negative. Positive charges repel positive charges, negative charges repel negative charges, and positive and negative charges attract. In other words, like charges repel each other and unlike charges attract each other. Normal atoms have an equal number of positive and negative charges, so usually atoms don't exert forces on one another. If you can somehow separate the charges, though, then you have either repulsive or attractive forces. If atoms are neutral, with equal numbers of positive and negative charges, then that explains why we don't see dramatic electric forces between atoms as an everyday occurrence. A neat solution, but then you have to figure out how those positive and negative charges are distributed in the atom. That leads us to the next section.

Caution: Do not use a "classroom demonstration" laser for the activity that follows. Anything stronger than a laser pen can be a danger to your eyes, and you should also be careful not to shine even a laser pen into anyone's eyes.

Even more things to do before you read more science stuff

For this section, you have a couple of alternatives. If you happen to have a laser pen or pointer, then you need that, a couple of sheets of cardboard, tape, and a few objects that are made of, at least in part, a reflective surface. A hand mirror will do, as will a watch with a glass front. If you don't have a laser pen,[2] you need

Figure 1.8

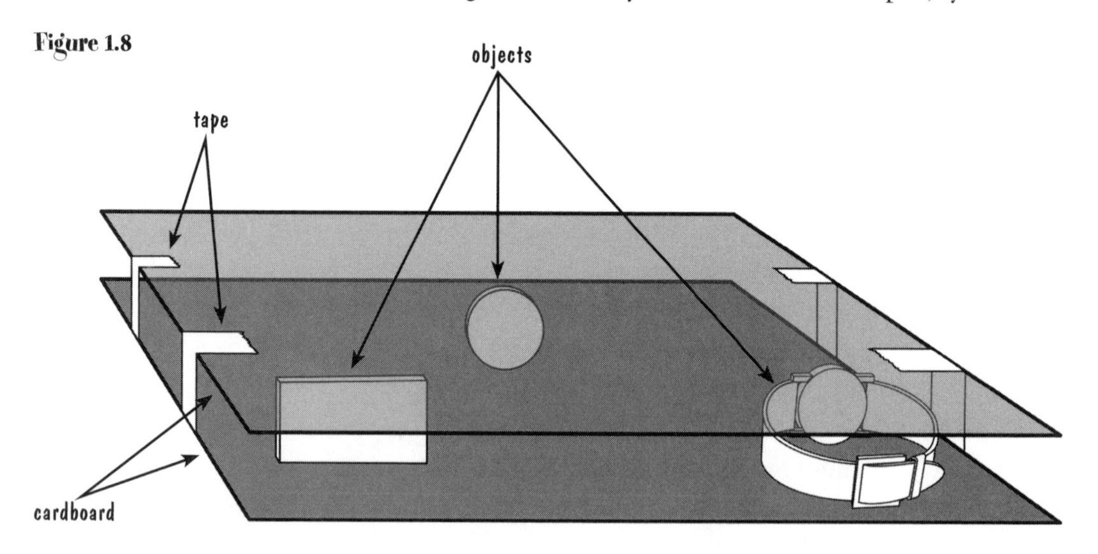

[2] In case you don't know, laser pens are relatively cheap, so you might want to go ahead and buy one. They're lots of fun to play with, and they're great for providing pets with exercise—just shine the pen on the floor or wall and watch as your pet chases the image all over the place.

a few marbles, the pieces of cardboard, the tape, and a few objects that are less breakable than a mirror or a watch.

In what follows, I'll describe what to do if you have a laser pen. After that, I'll explain how to adjust if you're using marbles instead of a laser pen. Take your mirror or watch and two pieces of cardboard, and arrange them as shown in Figure 1.8. Use tape to secure things as best as you can. This doesn't have to be a solid structure, but it should at least hold together for a while.

Now look from above and shine your laser pen in between the cardboard pieces from the side. By having a wall on one side of the apparatus or by placing a piece of paper on one side, you should be able to tell when the light shines straight through and when it bounces off something. With careful observation, you should be able to tell exactly where the light goes when it bounces off something. Take a look at Figure 1.9.

Figure 1.9

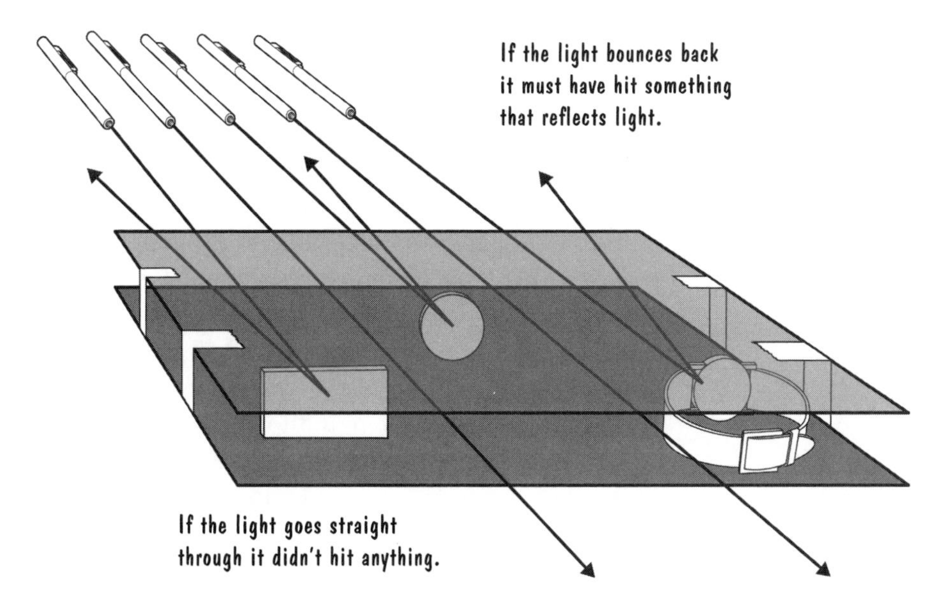

In this case, you know what's under the cardboard, so you can make sense of where the light goes. But what if you didn't know what was under the cardboard? Could you, with enough observations, get an idea of what was under the cardboard, or at least many of its features? Sure you could. It would take patience, but you could do it. To test that, have a friend tape a different object under the cardboard. Using just the laser and what happens to

SCI LINKS. THE WORLD'S A CLICK AWAY

Topic: Atomic Structures

Go to: *www.scilinks.org*

Code: CB003

the laser when you shine it in from the side, see if you can guess what's hidden under the cardboard. Yes, this will take a bit of time.

For those without laser pens. If you don't have a laser pen, you're going to have to shine something else on the object under the cardboard, something like marbles. By throwing marbles from the side, you can get some idea of what might be under the cardboard. They'll bounce off of something hard or something that's elastic. They'll just move on through if they don't hit anything.

Of course, this is more difficult than using a laser pen, so if you can spring about ten bucks for a laser pen, do that.

Even more science stuff

Topic: J. J. Thomson
Go to: *www.scilinks.org*
Code: CB004

Even though scientists figured out that there must be plus and minus charges in atoms,[3] they had no idea how those charges were arranged. One popular model was proposed by J. J. Thomson in the nineteenth century. He suggested that the positive and negative charges were evenly distributed around the atom, as shown in Figure 1.10.

Other models were proposed, and eventually people figured out a way to decide what might be inside an atom—shoot things at it! In an experiment that was a lot like what you did with your pen or marbles and the cardboard, a scientist named Ernest Rutherford did just that. He shot tiny things called alpha particles at a thin sheet of gold foil. Alpha particles have a positive charge, and if atoms look like Thomson's model, you would expect them to fly on through the gold foil, with maybe slight deflections when they get near the mixture of positive and negative charges in the gold atoms. Well, that's not what happened. Many alpha particles did go right on through, but some bounced straight back or nearly straight back, as if they had hit a brick wall. It was like a laser light hitting a piece of mirror or a marble hitting a solid object in the activity you did. Look at Figure 1.11 (p. 11).

Figure 1.10

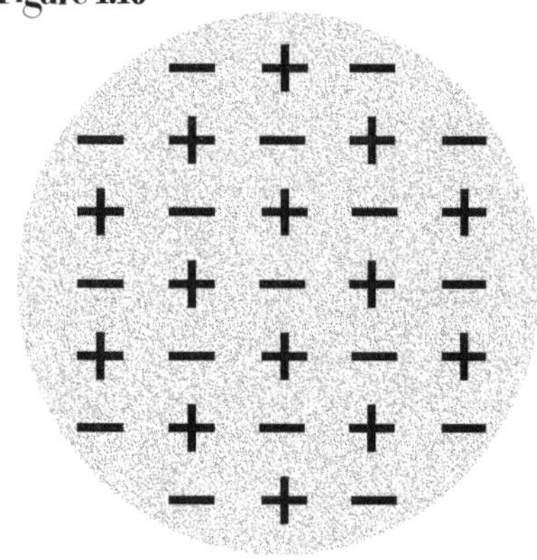

Thomson's model of the atom: evenly dispersed negative and positive charges.

[3] The labels of plus and minus are rather arbitrary. As explained in the *Stop Faking It!* book on electricity and magnetism, these might as well be called red and blue things that are inside atoms. It's all part of model building.

Figure 1.11

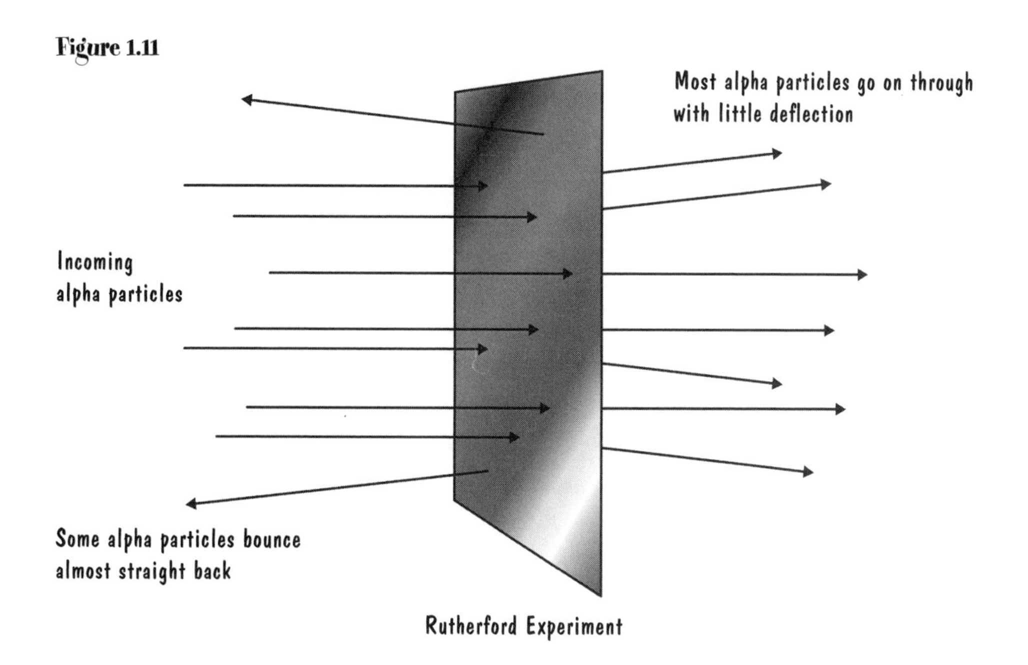

Rutherford Experiment

The only explanation Rutherford could come up with was that all of the positive charge in the gold atoms was concentrated somewhere in the atom. Only a high concentration of positive charge could repel the positively-charged alpha particles enough to send them back the way they came. He further reasoned that the concentrated positive charge had to be at the center of the atom. If it weren't, then the uneven distribution of concentrated positive charges would push on one another until they attained a symmetrical distribution, namely at the center of each atom. So, from Rutherford's time on, the accepted model of an atom was that there was a concentrated positive charge at the center, surrounded by negative charges in many other places around the atom, as shown in Figure 1.12 (p. 12). Later models would show that most of an atom is empty space, with neither positive nor negative charges. And for the record, we have a name for those negative charges in an atom—they're called **electrons**.

Topic: Rutherford Model of the Atom
Go to: *www.scilinks.org*
Code: CB005

Topic: The Electron
Go to: *www.scilinks.org*
Code: CB006

And even more things to do before you read even more science stuff

This is a pretty short section of the chapter. Here's a challenge for you. Assume that all matter is made up of atoms that have a positive charge at the center and a negative charge dispersed around it. Also assume that there are many kinds

Figure 1.12

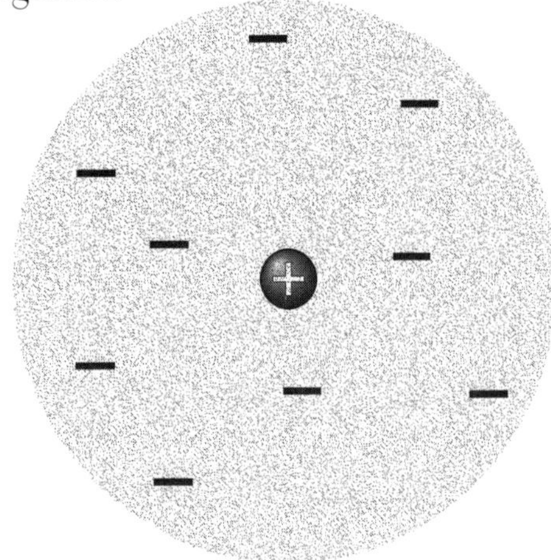

A new model of the atom.

of atoms in the universe. Using this model, see if you can come up with explanations for the activities I had you do in the first part of this chapter (baking soda and vinegar, burning a match, etc.). Try to explain all your observations, including any weight loss that occurred. No fair using any chemistry you might already know. Just use the model of the atom I've given you.

And even more science stuff

This is an even shorter section of the chapter. All I want you to do is realize that you couldn't get very far in your explanations in the previous section. In fact, you might say that the theory of earth, fire, water, and air is still better at explaining things than a model of atoms that consists of positive charges at the center and negative charges surrounding them. Of course, we're going to refine our model of the atom in future chapters, so you can guess which model—the model of Empedocles or the atomic model—wins out in the end.

Chapter Summary

- The notion that the universe is composed of elements—substances that cannot be simplified—has been around for centuries.
- The early Greek philosophers proposed that the universe was composed of four basic elements—earth, water, air, and fire. This theory can be used to explain a large number of observations.
- Soon after the theory of four elements, other philosophers proposed that the universe is composed of indivisible objects called atoms, and that elements in the universe are composed of only one kind of atom.
- Early experiments with electricity led scientists to propose the existence of two kinds of electric charges—positive and negative. This idea became part of models of the atom.
- Rutherford's experiment demonstrated that atoms have a concentrated positive charge surrounded by a disperse negative charge.
- At the end of this chapter, we do not have a complete model of the atom.

Applications

1. In describing motion, the Greek philosopher Aristotle relied heavily on the four-element theory of Empedocles. With that theory, Aristotle didn't need gravity or friction to explain why things did what they did. Objects fell to Earth because they were made primarily of earth and sought their proper place in the universe, not because of some unseen force of gravity. Objects similarly came to rest when you moved them across a surface, not because of a force of friction, but because once again they assumed their proper place in the universe. Not a bad idea actually, because Einstein's theory of general relativity, which is the currently accepted theory of forces and motion in the universe, uses this idea.

2. Another early Greek philosopher named Lucretius used a primitive model of atoms to explain how we smell things. He proposed that substances gave off atoms of a particular size and shape. These atoms would enter the pores of the human body only if their size and shape matched the size and shape of the pores. Different pores in the nose accepted different atoms, and through this mechanism humans identified different smells. This proposed process is remarkably similar to our currently accepted theory of how people smell things. Instead of pores, we refer to different "receptor sites" in the nasal membrane that connect with different-shaped molecules. More on this in the Applications section of Chapter 6.

Better Models

At the end of the first chapter, you realized, I hope, that the model of an atom that we have so far (Rutherford's model of a concentrated positive nucleus with negative charges around it) doesn't go very far in helping us explain observations. So, why not try to make that model better?

Things to do before you read the science stuff

Gather together a lightbulb from a flashlight, a 1.5-volt battery, three wires about 30 centimeters in length (stripped at the ends if they're insulated), and some masking

tape. You're going to put together the contraption shown in Figure 2.1. If you're up for a trip to the hardware store or RadioShack, take this book and show the people there what you want to build, and they'll be able to hook you up with everything you need, including a base for the bulb and a holder for the battery. The base and the holder aren't necessary, but they're more convenient than using masking tape.

Figure 2.1

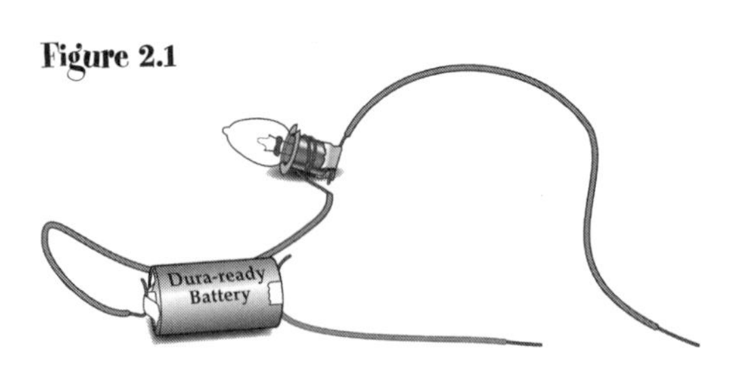

Touch the free ends of the wires together. The bulb should light. If not, take another look at the drawing and make sure your connections are correct. If the bulb does light, you're ready to go. Touch the free ends of the wires to different parts of various objects and see whether or not the bulb lights. Check out Figure 2.2 for a few examples.

Figure 2.2

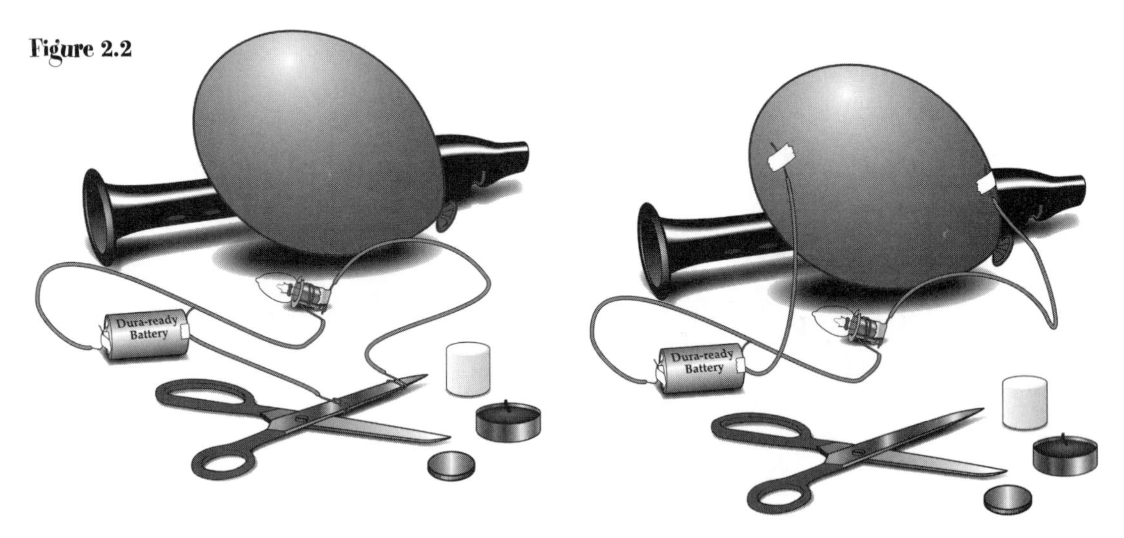

If the bulb lights, we might guess that "something" is flowing through the wires, the object, and the lightbulb. We call that something electricity. If the lightbulb doesn't light, then we can conclude that electricity does not flow through the object we're testing. The term we use is that the object does or does not **conduct** electricity. Once you know that, you can figure out what kinds of *materials,* rather than *objects,* conduct electricity. And once you think you have a grip on what materials conduct electricity, see if you can figure out what the atoms in a conductor of electricity look like compared to the atoms in a non-conductor of electricity. Not an easy task unless you already know the textbook answer.

SCILINKS
THE WORLD'S A CLICK AWAY
Topic: Conduction
Go to: *www.scilinks.org*
Code: CB007

Time for a field trip. Look at your surroundings and see what materials people use to make up radio and television antennas.

These can range from the antenna on a portable radio to the dish antenna that provides your satellite TV to those big towers on buildings and on mountains that broadcast the signals from local radio and TV stations. Of course, if you have a good memory this can be a field trip of the mind. Just remember what materials all those antennas were made of.

Finally, notice what you have to do to change channels on either a television or radio. Yeah, yeah, you turn a knob or click the remote, but what do you suppose turning the knob or clicking the remote *does*? How does that help you view different TV stations or listen to different radio stations?

The science stuff

Before addressing the activities of the previous section, let's talk about what causes a lightbulb to light. The presently accepted explanation is that electrons—those negatively-charged particles that surround the positively-charged nucleus in our current model of the atom—are moving through the lightbulb filament, causing it to produce light. So, in order to make sense of the previous activities, it helps to know that lightbulbs light because of electrons in motion.

As you went around touching the free ends of the wires to various objects, you probably weren't surprised. Everything made of metal conducted electricity (meaning electrons moved through the material) and lit the bulb; and things made of nonmetals, such as wood, plastic, and cloth, did not conduct electricity and did not light the bulb. Probably not a big surprise, because most of us learn the difference between conductors (materials that conduct an electric current) and insulators (materials that don't conduct an electric current) by the time we're in high school. What's more interesting for our present purpose is how this knowledge contributes to a better understanding of what atoms look like. If metals conduct electricity well, that must mean that at least some of the electrons in the atoms of metals are more or less free to move around. If nonmetals do not conduct electricity well, then that must mean that electrons in the atoms of nonmetals are *not* free to move around, at least not much. Chemists speak a great deal of metals and nonmetals, as this clearly is a major characteristic that distinguishes some elements from others. A bit more on this when we begin discussing the Periodic Table in the next chapter.

Okay, on to various kinds of antennas. What are they made of? Metal, of course. Your satellite dish for your TV is metal, the small antenna on your portable radio is metal, and those large antennas that broadcast signals from TV and radio stations are made of metal. Now, what do we know about metals? All we know at this point is that electrons can move around in metals. That knowledge is key, though, for understanding what happens when materials give off or absorb energy, as in when an antenna sends out radio and TV signals or takes in radio or TV signals.

If you can move electrons back and forth, or up and down, really fast, then you can create radio or TV signals. For future reference, we'll refer to that movement of electrons up and down or back and forth as **oscillations**. When electrons oscillate, they give off energy. The rate at which you oscillate those electrons, known as the **frequency** of the oscillation, determines how much energy they give off in the form of what are called **electromagnetic waves**.[1]

Okay, so electrons can oscillate up and down or back and forth and produce electromagnetic waves. It turns out that the reverse also works. Electromagnetic waves can cause electrons to move up and down or back and forth. Figure 2.3 illustrates how you can use oscillating electrons to produce electromagnetic waves and then use those electromagnetic waves to cause other electrons to oscillate.

Figure 2.3

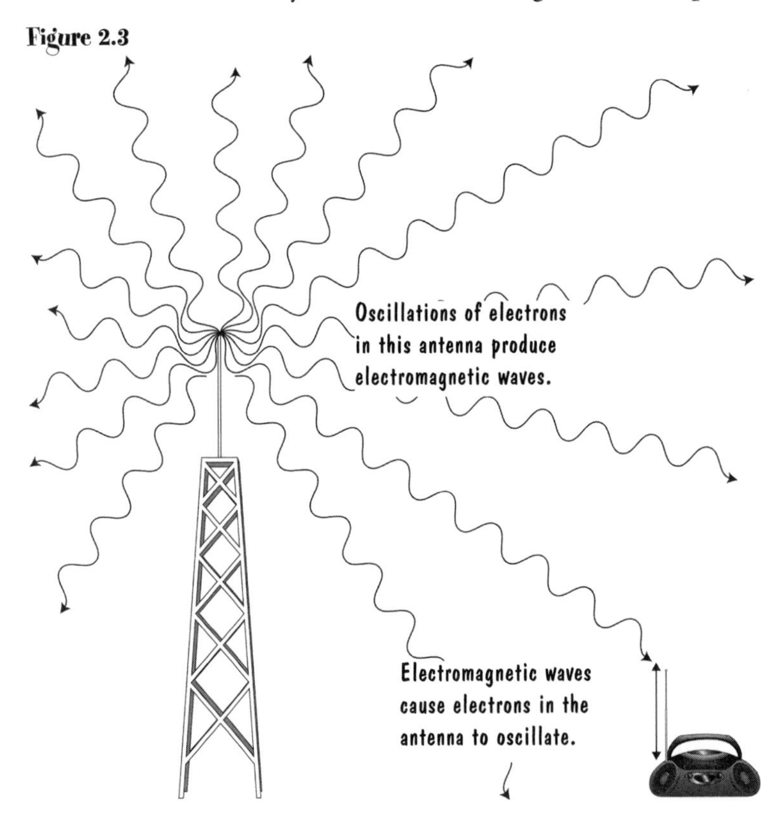

Figure 2.3 is the basic setup for transmission and reception of radio or television signals. Each radio or television station oscillates its electrons at a certain frequency, and each of those signals is causing the electrons in your antenna to oscillate at a different frequency, all at once. All you have to do is tune your set to different frequencies to pick up those different signals.

[1] I should note that I'm doing something here that doesn't do much in the way of fostering understanding, which is to introduce new science terms without a lot of explanation. I normally criticize people for doing that. In this case, however, the focus is on creating models of what is going on with atoms, rather than on the particulars of oscillations and waves. If you want to read more about these things, feel free to check out the *Stop Faking It!* books on light and sound. And yes, that was a shameless promotion.

About now you're probably wondering what in the world radio and TV signals have to do with a model of atoms. Nothing directly, actually, but I'm setting you up to understand how we can use light waves to discover a lot about atoms. Just be patient until the next explanation section, okay?

Figure 2.4

The antenna receives radio signals of many different frequencies.

More things to do before you read more science stuff

For this next activity, you need to get something called a **diffraction grating**. If you happen to own the *Stop Faking It!* book on light, then you already have a diffraction grating called a "rainbow peephole" that is included with that book. If you don't own that book, then check in the science supplies in your school. If you can't locate a diffraction grating there, find a science store or contact the company Rainbow Symphony[2] and they'll hook you up. Once you have a diffraction grating, turn on a bare lightbulb, darken the rest of the room, and look at the bulb through the grating. You should see all the colors of the rainbow, spread out to the right and left, and possibly in other directions depending on the kind of diffraction grating you have. See Figure 2.5.

Topic: Diffraction

Go to: *www.scilinks.org*

Code: CB008

Figure 2.5

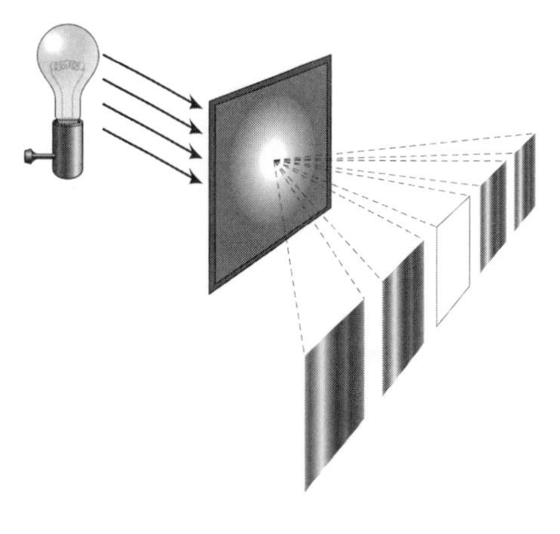

A diffraction grating separates white light into a number of rainbows.

Now take your diffraction grating out at night (it's good for them to get out on the town once in a while) and find street lamps and/or neon signs. Look at these lights through the diffraction grating, and notice any difference between what you see now and what you see with a regular lightbulb. For the record, you should only see certain colors rather than all the colors of the rainbow.

Head back inside and get yourself a rope, a string, or a Slinky. Tie or otherwise attach one end to something solid. Holding the free end,

[2] Just go to *www.rainbowsymphony.com* on the internet. That site has lots of inexpensive optics items that are of good quality.

begin moving the string[3] up and down slowly. Gradually increase the frequency of your oscillations (how fast you move the end up and down) until you get the pattern shown in Figure 2.6.

Figure 2.6

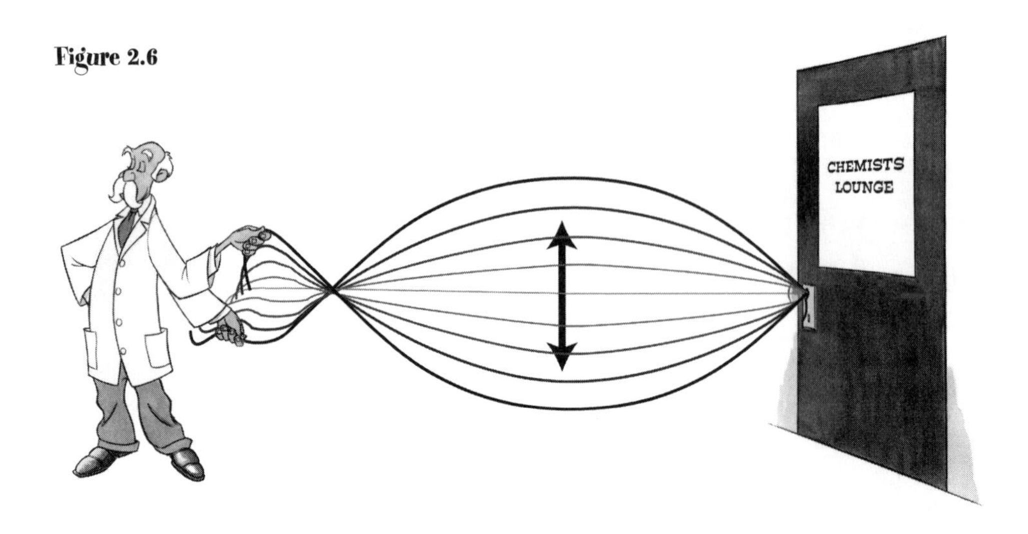

Topic: Wavelength
Go to: *www.scilinks.org*
Code: CB009

Gradually increase the frequency of your oscillations again, until you can get other neat patterns in the string, such as those in Figure 2.7 (p. 21). These patterns are known as **standing wave** patterns. The reason they're called standing waves, as in "stationary waves," is that the overall pattern keeps its shape even though there's lots of motion going on. As you're creating these standing waves, think about how much energy you expend to make each pattern. Is there a correlation between the energy expended and the frequency with which you move the end up and down?

Next change how much of the string you're moving up and down by moving your hand from the free end toward the center of the string. With this shorter string, try and reproduce the standing wave patterns you got previously. Do the patterns happen at the same frequencies (how fast you move the end up and down) as before? Your answer to this should be no. If not, reread what you're supposed to be doing![4]

[3] For simplicity, I'll assume you have a string from here on. What I say also applies to a rope or a Slinky.

[4] For those of you familiar with how scientists describe waves, you will note that as the frequencies of the waves change, so do their wavelengths. We'll stick to using the term *frequencies* in our discussion.

Figure 2.7

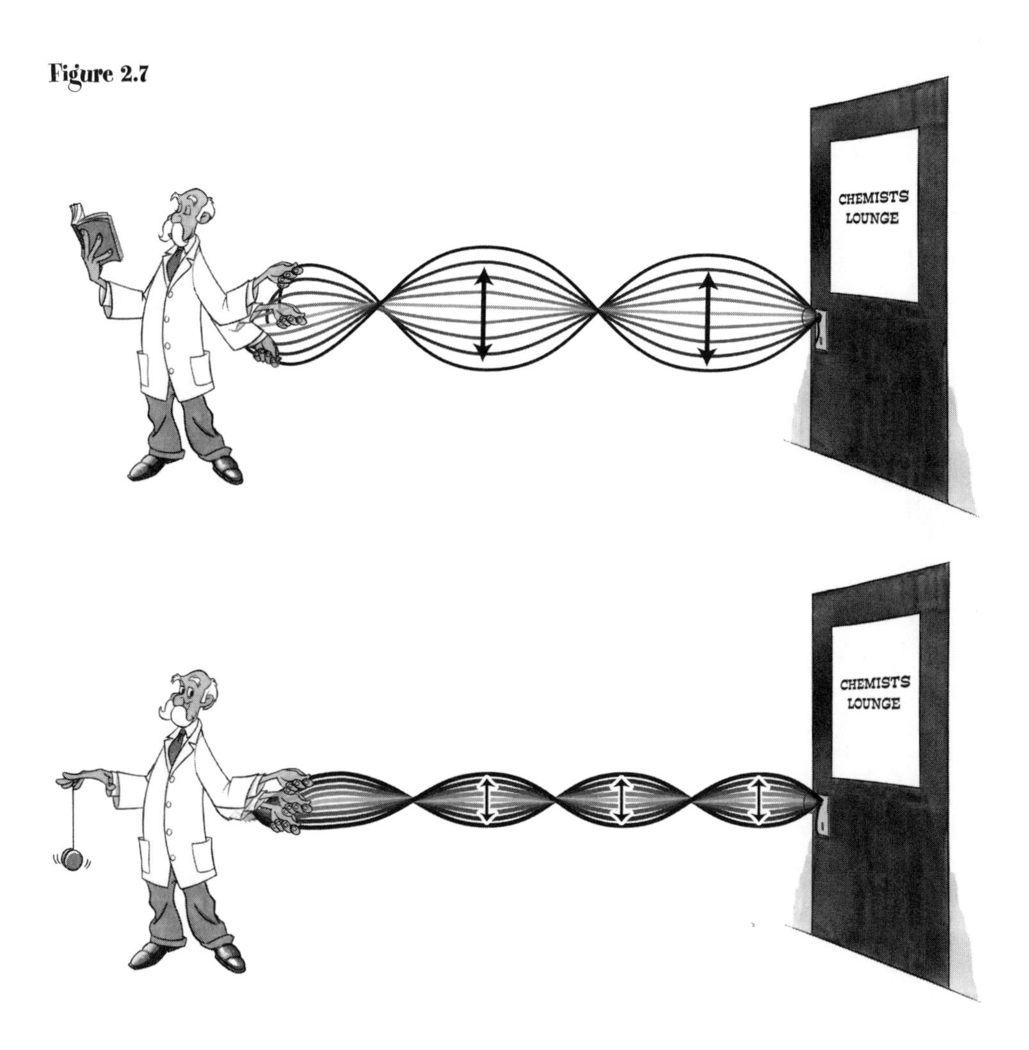

More science stuff

First, let me tell you what diffraction gratings do. They separate light and other electromagnetic waves out into different frequencies. Yes, light is an electromagnetic wave, exactly like radio waves but with different, higher, frequencies. Prisms also do this, but not as neatly and precisely as diffraction gratings.[5] A diffraction grating actually sends light waves to different positions depending on the frequency of light. Different frequencies show up in different positions. Because different frequencies of light correspond to different colors, the different colors appear at different positions when you look at white light through a diffraction grating.

[5] To learn more about diffraction gratings, check out—you guessed it—the *Stop Faking It!* book on light.

Because you can see rainbows of color when you look at a lightbulb through a diffraction grating, that means that the atoms in the bulb filament must be emitting electromagnetic waves that correspond to the frequencies of light waves. That presents a problem for our model of the atom, though. We learned earlier that you can create radio waves just by making electrons oscillate in a metal. Fine, but you can produce light simply by heating something up, without any mechanism for causing electrons to oscillate. Also, you can create light by heating up *non*metals. We saw earlier that nonmetals don't conduct electricity, so the electrons in nonmetals are not free to move all over the place. To complicate things even more, we have discovered that heated gases such as mercury and neon only emit certain colors of light. When you look through a diffraction grating at the light from a street lamp or a neon sign, you only see certain colors of light. Street lamps usually contain mercury gas, and only mercury gas. When you look at a street lamp, you are looking at the light emitted by only one kind of atom, namely mercury. Similarly, a neon light produces only light emitted by the gas neon. Different atoms, when excited by an electric current or by some other input of energy, emit definite patterns of light frequencies. In fact, the frequencies of emitted light serve as fingerprints that one can use to identify elements. Cool, but we have to explain all this.

Let's summarize our situation:

1. Making electrons oscillate in metals produces electromagnetic waves.
2. Light waves are electromagnetic waves.
3. Even nonmetals can produce light when heated, and we don't have a ready mechanism for oscillation of electrons in nonmetals.
4. The light produced by atoms appears to have only certain frequencies and energies.
5. Clearly we need a new mechanism besides oscillation of electrons in metals for the production of light by atoms.

The answer to our problem will seem a bit bizarre, but it's a model that works quite well. First, it turns out that electrons, just like all other things in the universe, can be considered to be a compilation of waves. You can consider light to be particles or waves, and you can even consider humans to be a collection of waves. If you've ever heard the term *wave-particle duality,* that's what I'm talking about. So anyway, what we are doing is confining electrons to atoms, which amounts to confining a collection of waves to atoms.

Here is where the standing waves on a string come into play. If you'll recall, a given length of string at a given tightness only seems to "like" certain frequencies. When you hit one of those frequencies, the oscillation of the string goes wild and you get those cool standing waves. The lesson we can learn from this is that when you confine oscillations, only certain frequencies are accepted by the confining situation. If you change the situation, such as when you change the length of your string, you change which standing wave frequencies fit. One last thing on those standing waves. It should make sense that different frequencies are associated with different energies. Just think about how fast you had to move the string up and down. Higher frequencies required faster oscillations, and more energy from you. So, confining waves on a string means that only certain frequencies, and thus certain energies, are allowed.

Let's apply this information about confined waves to what might be going on inside atoms. Electrons are bound by electric forces to atoms. Electrons can also be thought of as a collection of waves, or oscillations. When you confine oscillations, only certain frequencies and energies flourish. Also, those certain frequencies depend on the confining conditions. Therefore, we might expect that electrons confined within atoms only oscillate at certain frequencies, and those frequencies will be different from one kind of atom to another. And yes, that's what happens. The waves that compose electrons are only allowed to have certain values in a particular atom, and thus electrons in atoms are only allowed to have certain energies.

At this point it is tempting to think that the energies and frequencies of the electron waves in atoms correspond to the energies and frequencies of light emitted by atoms. Tempting, but wrong.

Physicist Niels Bohr to the rescue. He developed a model of the atom that had electrons residing in certain **energy levels**, which by no small coincidence happen to correspond to the energy levels we predict from confining electron waves to atoms. When they jump from a higher energy level to a lower energy level, they radiate light with an energy that corresponds to the difference in energy between the two levels. When atoms absorb light of just the right energy and frequency, electrons jump from a lower energy level to a higher energy level. Take a look at Figure 2.8 (p. 24). The key here is that the frequency and the energy of the light emitted or absorbed correspond to the frequency and energy of the *difference between energy levels,* not the frequency and energy of the electron waves themselves.

Topic: Energy Levels
Go to: *www.scilinks.org*
Code: CB010

So:

1. electrons can be thought of as waves.
2. confining electrons to atoms means the electron waves can only have certain energies and frequencies.

Figure 2.8

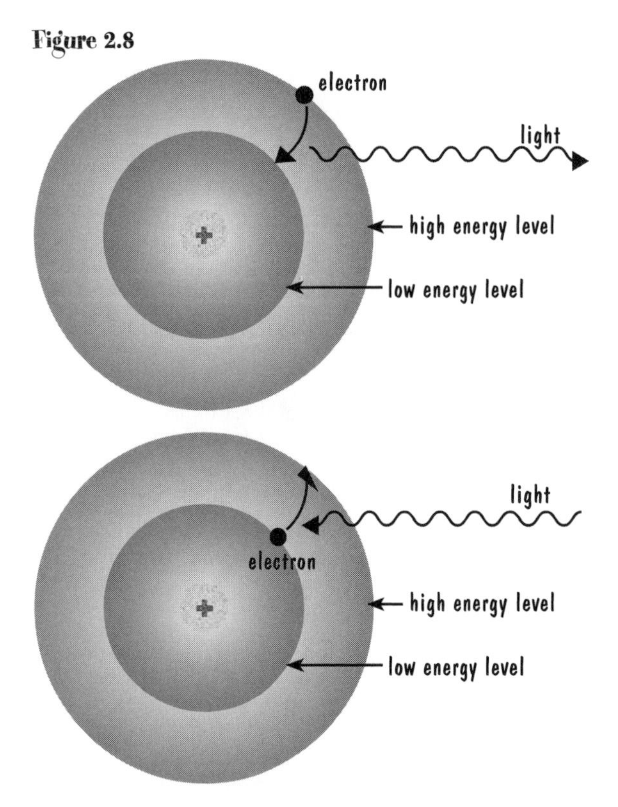

3. atoms emit light and absorb light that has energies and frequencies that correspond to the differences between electron energy levels.

So, we now have a positively-charged nucleus at the center of an atom, with electrons residing in only certain energy levels around that nucleus. We also know that electrons can jump between energy levels. This model helps explain how atoms interact with light and emit light, but it still isn't of much use in explaining basic chemical reactions. You're going to have to wait another chapter before we get to an even better model.

Where we're headed. Just as a reminder, we are gradually building better and better models of what atoms look like. We began with the Greeks and shifted gears to a primitive model of the atom. Then Rutherford helped us add a positive nucleus at the center. With Bohr's help, we now have a rough idea of what the electrons in atoms are doing. They reside in definite energy levels that are atom-specific, and when they change energy levels they can emit or absorb light.

Even more things to do before you read even more science stuff

Find a large rock and a small rock, or just one thing that's heavy and one that's not very heavy. Put each object on a flat surface and push both along with your hand. Which one is easier to push? Pretty simple, huh?

Get a weak magnet such as a refrigerator magnet or one you can find at a hobby or craft store. You don't want anything close to a strong magnet, so even a regular bar magnet or horseshoe magnet is too strong. Turn on a television set and *slowly* bring your cheap magnet near the screen, moving it from side to side as you do this. You should notice distortion on the screen.[6] Once you see the distortion, don't bring the magnet any closer to the screen, or you might permanently alter the picture quality in a given region of the screen.

Even more science stuff

The large rock, providing it was heavier than the small rock (or other object), was more difficult to push, right? An object's **mass** (related to its weight, but not the same thing) is a measure of how difficult it is to change the object's motion. (See Figure 2.9.) So, that heavier rock had more mass and was more difficult to push.[7] Simple conclusion: You can learn a lot about an object's mass by trying to change its motion. In fact, you can get an *exact* determination of an object's mass this way.

SCILINKS. THE WORLD'S A CLICK AWAY
Topic: Mass
Go to: *www.scilinks.org*
Code: CB011

On to the TV screen and the magnet. What you might not know is that televisions work by firing electrons at the front screen. You putting a magnet near the screen demonstrates that you can deflect the motion of electrons (they hit in the wrong spot, causing distortion) by putting a magnet near them. In fact, all moving charged particles, not just electrons, are pushed in a circular path by a magnetic field produced by a magnet (Figure 2.10).

Figure 2.9

The harder it is to change an object's motion, the more mass it has.

Figure 2.10

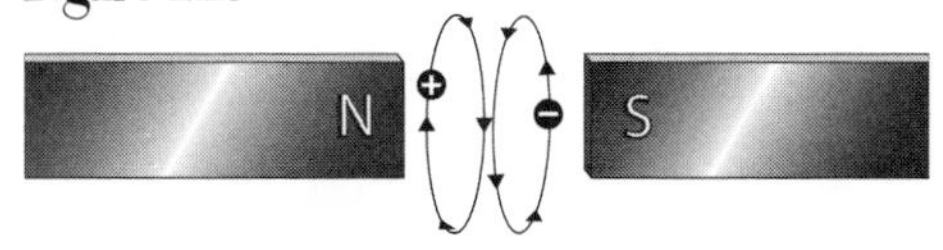

Moving charged particles are deflected into a circular path by a magnetic field.

[6] It is important that you use a *weak* magnet. Test your magnet on the fridge or other metal object first. If there is a strong attraction, find a weaker magnet. If you have any doubts, do this with an old computer monitor or television you're about to give to charity.

[7] This is at best a brief introduction to forces and mass. For more information, consult your favorite book that deals with force and motion. Gee, I wonder if there's a book series that deals with this.

What's this have to do with our model of the atom? Well, we're about to find out how you can determine the charge on a proton and on an electron, and also determine the mass of each of these. Imagine that you have two plates that have opposite charges. You also have an atomizer (like a perfume sprayer) that can shoot very tiny drops of oil in between the plates. If these drops happen to pick up excess electrons when you spray them (and they do), then the charges on the plates will cause the tiny drops to move slowly from one plate to another. The setup is shown in Figure 2.11.

Figure 2.11

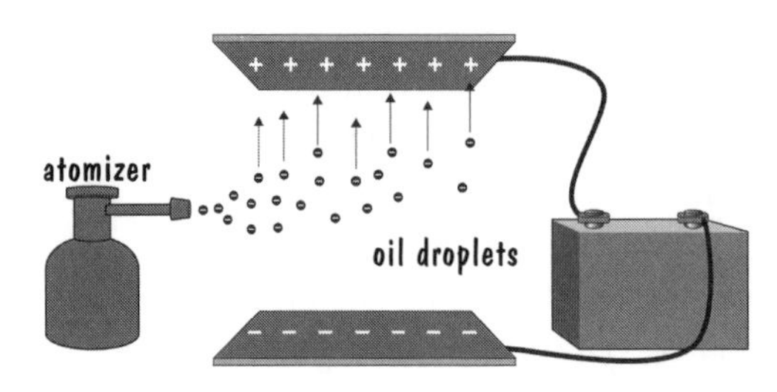

This setup is similar to a classic physics experiment known as the Millikan oil drop experiment. You don't suppose the guy who did it was named Millikan, do you? I won't go into a whole lot of detail on how to do it, but you can use this experiment to determine the charge on a single electron. For the record, the charge on an electron is 1.6×10^{-19} coulombs, which probably doesn't mean a whole lot to you, but at least now you know the number.

The next thing I need to explain is something called a **mass spectrometer**. What this device does is shoot charged particles into a magnetic field. They then move in a curved path with a radius that depends on both the mass of the particle and the charge of the particle. In fact, you can determine the value of the charge of the particle divided by the mass of the particle. And this is where our Millikan experiment comes in. If you know the charge on an electron and you know the value of the charge divided by the mass, you can determine the mass of an electron. Same thing goes for a proton.[8] Figure 2.12 (p. 27) shows a drawing of a mass spectrometer.

Topic: Mass Spectrometry
Go to: *www.scilinks.org*
Code: CB012

So where does this leave us? We know the charge on an electron, the charge on a proton (the same as the charge on an electron), the mass of an electron, and the mass of a proton. Turns out that the mass of a proton is *much* larger than the mass of an electron—about twenty thousand times as large. This last fact will be important in the next two sections, and the actual numbers are shown in Table 2.1 (p. 27). It will help if you know scientific notation.

[8] You might wonder how one can isolate electrons or protons. It turns out to be a relatively easy process if you start with hydrogen, which is the simplest atom we know of.

Figure 2.12

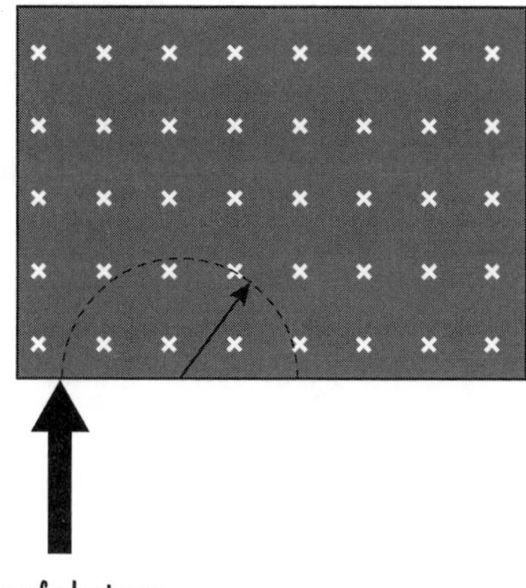

Table 2.1

Particle	Mass	Charge
proton	1.67×10^{-27} kilograms	$+ 1.6 \times 10^{-19}$ coulombs
electron	9.1×10^{-31} kilograms	$- 1.6 \times 10^{-19}$ coulombs

And even more things to do before you read even more science stuff

This next activity is a simulation of the kinds of things you can discover with enough experiments. The reason it's a simulation is because, frankly, you can't perform the necessary experiments with your basic household materials. To get the full impact of the simulation, it would be best if someone else read the rest of this section and set things up for you. If no one else is around, set it up and, at the appropriate time, pretend to be clueless as to what's happening.

Go to the store and get two or three different sizes of marshmallows. You'll need at least three of each size, and it's okay if all you have are regular marshmallows and mini-marshmallows. Also stop by the hardware store and get about a dozen small metal ball bearings.[9] Now, using something long and thin (I used one of those metal picks that come with nutcrackers) poke a narrow cavity in one end of two of each size marshmallow. Don't poke all the way through (see

[9] Ask at the hardware store if you don't know what these are. The helpful people at the helpful hardware store will know what you need.

Figure 2.13

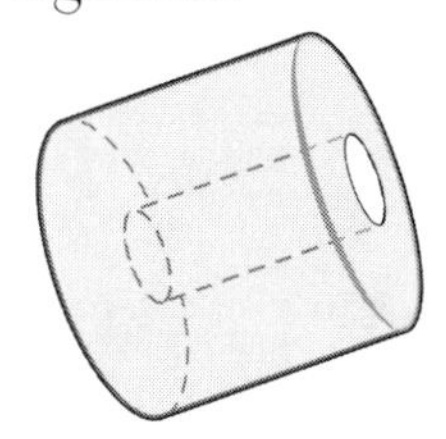

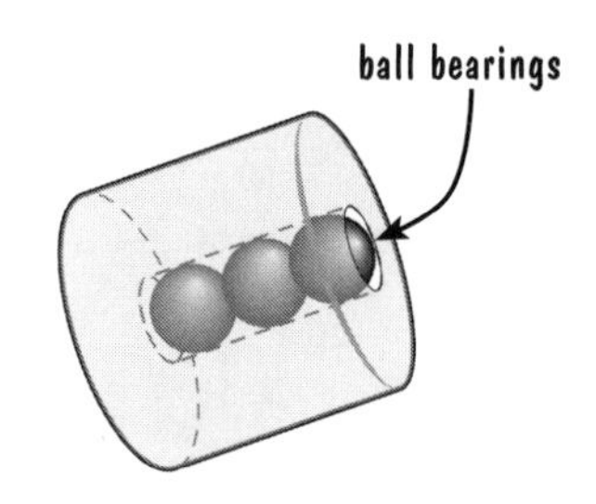

Figure 2.13). Then place one ball bearing in the cavity of one marshmallow and two or three ball bearings in the cavity of another marshmallow of the same size. Repeat with the other size(s) of marshmallows.

When you're done, you should have three of each size marshmallow, two of which contain ball bearings. Place them all cavity-side down, and you should have what appear to be almost identical marshmallows, as in Figure 2.14.

Time for your friend to present the marshmallows to you, or for you to pretend you don't know what you just did if you're by yourself. Look at the marshmallows. Does each one of the same size look pretty much the same? When you bring them close to one another, do any of the marshmallows seem to attract or repel other marshmallows? How do you suppose the number of protons and electrons in each marshmallow compare? Remember, protons are positively charged and electrons are negatively charged, and they have equal-sized charges. Finally, pick up each marshmallow, without turning it over, and compare. Any difference? Of course there's a difference. Some of the identical-looking marshmallows are heavier than others. If you didn't know how the marshmallows were prepared, how might you explain this difference in otherwise identical marshmallows?

Figure 2.14

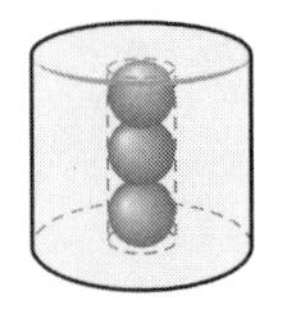

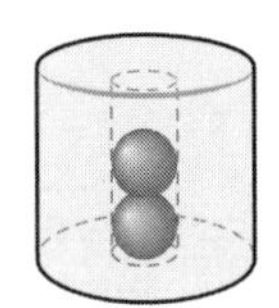

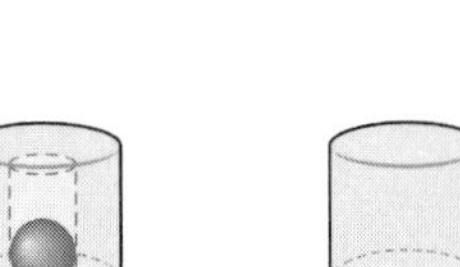

And even more science stuff

Yes, that was a silly activity to do. Try it with other people, and you'll get interesting answers for why some marshmallows weigh more than others. The obvious conclusion is that you have tampered with them, but beyond that, the answers vary. My son thought I had used different marshmallow material in different marshmallows, and my wife thought I had somehow shrunk or enlarged some of the marshmallows. Didn't take them too long, though, to figure out that there was weight added to the marshmallows, without otherwise changing the characteristics of the marshmallows.

The activity might be silly, but it's analogous to a dilemma that faced scientists years ago. They had figured out that different elements had different numbers of protons and electrons, because they had different masses and different properties. They also discovered, though, that collections of atoms that had similar properties had masses that didn't go along with a specific number of protons and electrons. It was like expecting all marshmallows that had identical properties to weigh the same, but finding out they didn't. What's the explanation with the marshmallows? Simply that we have added extra mass (and weight) that doesn't affect the outward properties of the marshmallow. The explanation for atoms was similar. The atoms contained a bunch of extra mass that didn't affect the properties all that much. This led to the invention of things called neutrons. Neutrons have about the same mass as protons, but no charge.

Figure 2.15

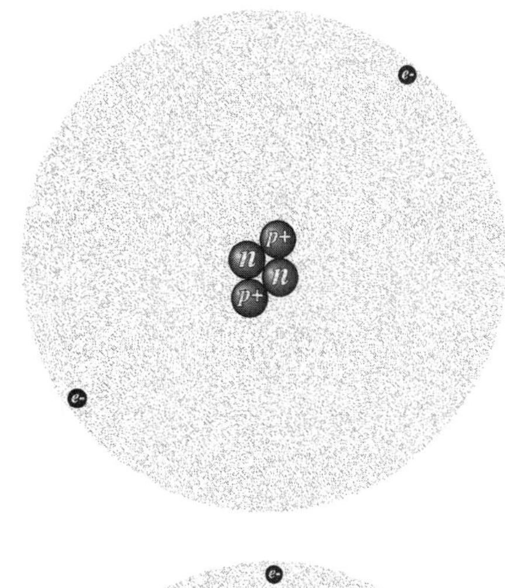

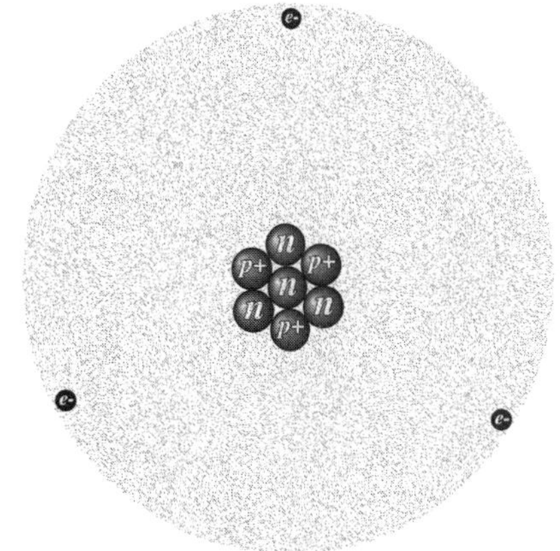

With no charge on them, it makes sense that you could add extra neutrons to an atom without changing its properties very much. You don't alter the charge of the nucleus, so that doesn't alter the overall charge of the atom, which as you might recall is normally neutral. If you don't alter the overall charge, then you don't add any electric forces, which certainly could be considered a change in the properties of an atom.

Anyway, here's the picture we end up with. An atom of any substance has protons and possibly neutrons in the nucleus. Sometimes the number of protons and neutrons is the same, but there often are more neutrons than protons. What constitutes "normal" for the number of protons and neutrons in an atom varies from atom to atom. Atoms with more neutrons that what is "normal" are called **isotopes**. Figure 2.15 gives you an idea of our current picture of a couple of atoms. And by current, I mean our current picture in this book, not the presently accepted picture of atoms according to scientists.

Topic: Isotopes
Go to: *www.scilinks.org*
Code: CB013

We should also add neutrons to our list of particles that make up atoms. That's shown in Table 2.2 (p. 30).

Table 2.2

Particle	Mass	Charge
proton	1.67×10^{-27} kilograms	$+ 1.6 \times 10^{-19}$ coulombs
electron	9.1×10^{-31} kilograms	$- 1.6 \times 10^{-19}$ coulombs
neutron	1.66×10^{-27} kilograms	0 coulombs

Before moving on to the next chapter, I want to say something more about my approach in this book. Sometimes I follow the historical development of chemistry, and sometimes not. When I don't follow the historical development, the reason is that I want to help you understand how a complete understanding of atoms and how they behave explains all sorts of chemical reactions. In my mind, that includes an understanding of the basics of how atoms are put together. Better that than memorizing what happens in those reactions.

Keep in mind something I say in almost all of my books, which is that all of science is made up. No one has ever seen a proton, neutron, or electron. They might be real, and they might not. The important thing is not that these made-up things are real, but that they help explain observations and predict new observations. I try to help you understand *why* people think things like protons and electrons exist, but the bottom-line evaluation is whether or not our model explains observations.

Chapter Summary

- Some materials, known as metals, conduct electricity and some materials, known as nonmetals, do not conduct electricity.
- The conduction of electricity is the movement of electrons through a substance.[10]
- Causing electrons to move up and down or side to side (causing them to oscillate) produces electromagnetic waves.
- It's not possible to make electrons oscillate fast enough to produce the electromagnetic waves that correspond to light. There must be some other mechanism.
- When you confine waves, you create a situation that is favorable only for certain wave frequencies and certain wave energies.

[10] This is not entirely true. The movement of *any* kind of charge, positive or negative, electron or not, is the conduction of electricity. For our purposes here, though, we can think of it as just the movement of electrons.

- We can model electrons, as well as all other particles, as waves.
- Confining electrons to atoms means electrons can only have certain frequencies and thus reside only in certain energy levels.
- When electrons jump from one energy level to another, they either emit or absorb light. That light has a frequency and energy that corresponds to the frequency and energy *difference* associated with the electron jump.
- With careful experimentation, you can determine the mass and the charge of various particles that make up atoms—protons, neutrons, and electrons.
- Atoms contain neutrons as well as protons in the nucleus. Neutrons don't have any charge, but have about the same mass as protons.
- Electrons don't have much mass compared to protons and neutrons.
- Atoms with "extra" neutrons are known as isotopes.

Applications

1. I mentioned that the pattern of light frequencies that an atom emits, known as its atomic spectrum, serves as a fingerprint for each atom. Astronomers use this information to find out lots about the universe. If you send the light from a distant star through a diffraction grating, you will see lots of overlapping atomic spectra. By sorting out these patterns, you can find out what elements are contained in that distant star. Similarly, we can find out what elements are present in the planets in our solar system. We get all this information without having to travel to these planets and stars.
2. You might be familiar with something called the Doppler effect with sound waves, where the sound you hear is a higher frequency if the sound source is moving toward you and a lower frequency if the sound source if moving away from you. Think about the sound that a train or a race car produces as it goes past you—higher pitch sounds until it reaches you and then lower pitch sounds as it moves away from you. When you look at the atomic spectra from distant stars and galaxies, they don't exactly match up with the spectra from elements on Earth. The reason for this is that these stars and galaxies are moving either toward us or away from us. Light waves experience a Doppler effect, just as sound waves do. With our knowledge of atomic spectra and the Doppler effect, scientists have concluded that the farther objects are from us in the universe, the faster they are moving away from us. This leads to the conclusion that the universe is expanding.
3. I implied in the chapter that adding neutrons to an atom doesn't change the properties of the atom very much. That's true only up to a point. Adding

extra neutrons can make the nucleus unstable, and in fact leads to atoms that are radioactive, and these atoms transform into more stable atoms along with the release of energy that can be either harmful or helpful to us humans.

4. The basic mechanism behind a mass spectrometer, which is that moving charged particles move in a circular path in a magnetic field, is also one of the main components of particle accelerators. These accelerators are often circular in shape. Powerful magnets in the accelerators cause charged particles to move in a large circular path as other components of the accelerator cause the particles to reach higher and higher speeds. Much of what we know about atoms results from using particle accelerators to slam fast-moving particles into other substances.

Periodicity

In thinking about a title for this chapter, the word *periodicity* came to mind. I was sure this had some kind of pop culture reference. After discussing this with my wife, we figured I was thinking of *synchronicity,* which is a reference to music by the band The Police. Looking the word *periodicity* up on the internet, I found that I was, in fact, a science geek and had not made a hip reference. Periodicity refers mainly to the Periodic Table, which is a focus of this chapter. No music, just science.

Anyway, it's time to refine our model of atoms to the point that we can understand the Periodic Table and a number of chemical reactions. To do that, it's necessary to understand how electrons behave and to understand a few things about the energy of electrons. Obviously, then, we'll have to begin by playing with Hot Wheels track. Okay, that's not obvious, but it's what we're going to do.

Before you begin this chapter, know that I realize the Periodic Table has many negative memories for people who have not done well in chemistry. It's actually a cool thing, though. Many scientists spent many a solitary moment in the lab and in the office trying to figure out how all the different atoms and elements related to one another. As a result, the Periodic Table has a whole bunch of information. It's a road map to the relationship between atoms, and it's nothing to be afraid of. My hope is that you will *understand* the reasoning behind the Periodic Table, and thus understand the patterns. If all you see are the patterns, then you have to memorize the patterns. If you understand the mechanism behind the patterns, you don't have to memorize anything. It will make sense.

Things to do before you read the science stuff

Grab a section or two of Hot Wheels track, hereinafter referred to as just "the track." If you don't have any such thing around, then you can use any kind of flexible surface on which you can roll marbles or cars. Foam pipe insulation from the hardware store works well (cut it in half lengthwise so you have a "half-pipe"). Tygon tubing that has an opening large enough for a ball bearing is another option. In addition to the track, you will need a marble, a ball bearing, or a toy car that rolls easily on the track.

Figure 3.1

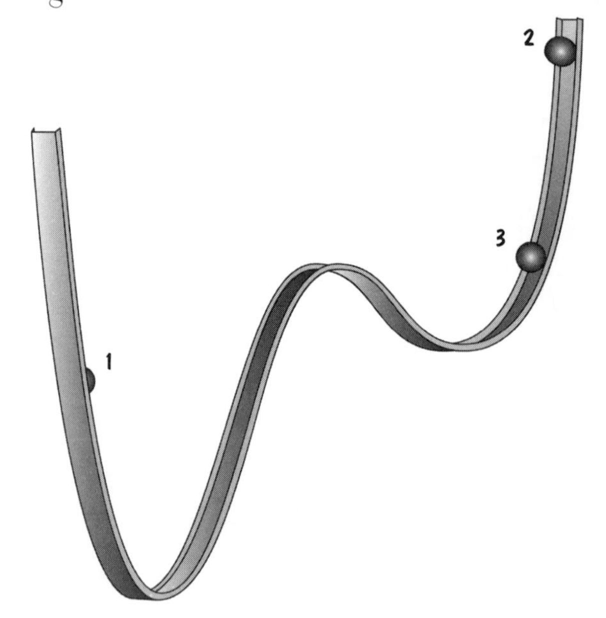

Before using the track, simply place your marble on the floor. Then raise it up a bit and let go. Then answer the following: In which position does the marble have more energy—on the floor or raised up above the floor? Once you've answered that, put your track in a shape similar to that shown in Figure 3.1. You might need a friend's help if your dexterity isn't superior.

Place the marble at point 1 on the track and let go. What does the marble do? Now place the marble at point 2 and let go, again observing what the marble does. Finally, place the marble at point 3, let go, and observe. Try to come up with a general rule that describes what the marble tends to do, no matter where on the track it's released.

Next get a couple of magnets. It would be nice to have strong bar magnets, but cheap refrigerator magnets will do. Arrange the magnets so they stick together (north pole to south pole if the poles are labeled on your magnets). Pull them apart slightly and let go. They snap back together, right? See Figure 3.2.

Figure 3.2

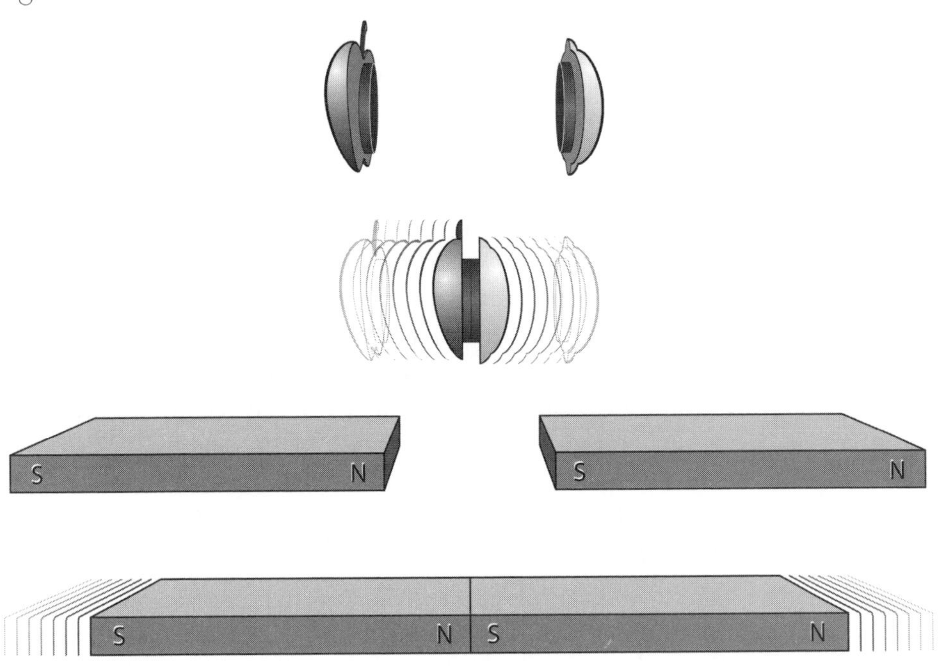

Now answer a question. In which position, pulled apart or stuck together, do the magnets have more energy? If you're up on your energy concepts, this is an easy question. If not, just give it a try and know I'll answer it for you in the next section.

Find a smooth surface and grab one of those metal ball bearings I had you use in Chapter 2. Arrange one of your magnets and the ball bearing so that when you let go of the ball bearing, it rolls toward the magnet. The weaker your magnet, the closer the two will have to be in order for the ball bearing to spontaneously move toward the magnet. Repeat this activity, but put an index card between the magnet and the ball bearing. What happens? See Figure 3.3 (p. 36).

Figure 3.3

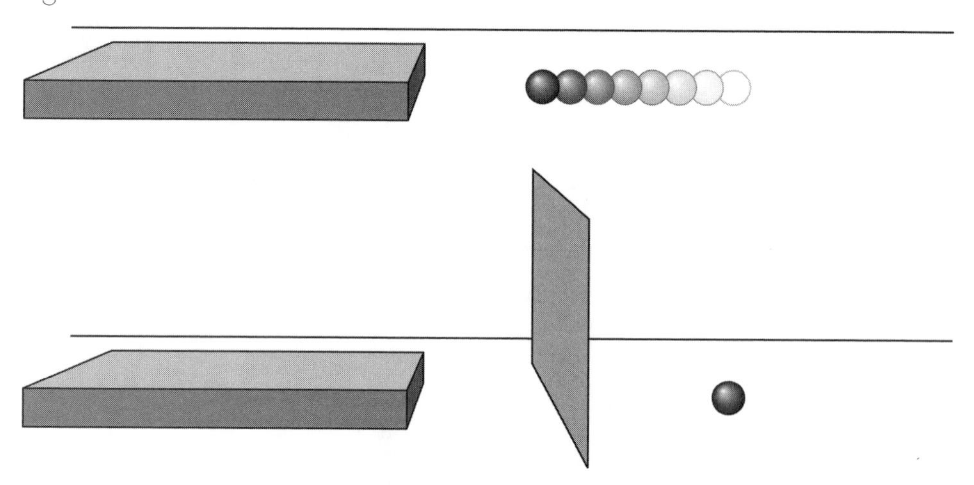

The science stuff

First I'm going to talk about the marble alone, when you have it on the floor and then raised above the floor. Before answering the energy question, I need to explain that any energy the marble has as a result of its position with respect to the floor is *shared* between the marble and the Earth. The gravitational pull of the Earth is what makes the marble fall to the floor, so the Earth has to be part of the picture.[1] That said, the energy between the marble and the Earth increases the more they're separated. We know this because of the energy that we can see when we let the marble go. If you drop a marble from a high height, it gets moving faster than if you drop it from a low height. If you let go of it while it's on the floor, it doesn't move at all. So, the greater the separation of the marble and the Earth, the greater the energy they have. See Figure 3.4.

Figure 3.4

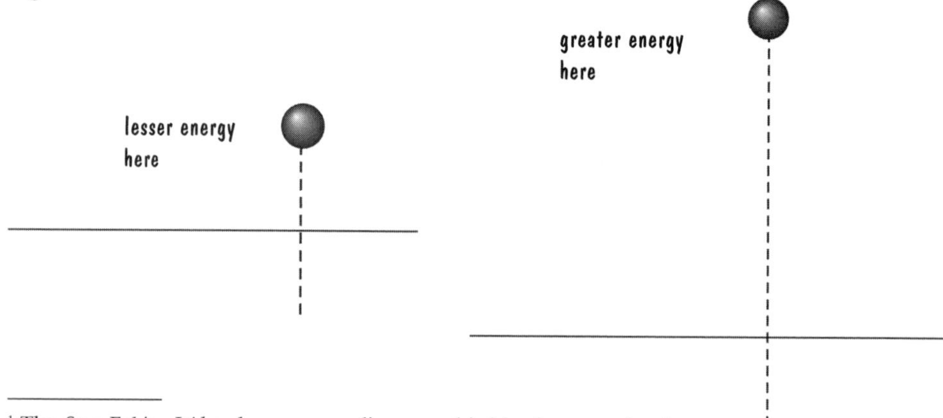

[1] The *Stop Faking It!* book on energy discusses this idea in more depth.

The situation is almost identical with two magnets. The farther apart you place them, the faster they're moving when they slam back together. So, the farther apart they are, the more energy they possess. Of course, there's a separation beyond which the magnets won't go back together, but the reason for that has to do with friction and other forces affecting things; if the two magnets were the only objects in the universe, they would eventually slam together no matter how much they were separated.

On to the track and the marble. What rule did you use to describe what the marble does? You might have said something like, "The marble tends to go downward." That's a good rule. Here's one you probably didn't state: The marble tends to move to a position where the energy shared by it and the Earth is the lowest. Yeah, didn't think you'd come up with that one. This rule works pretty well, though. The closer the marble is to the Earth, the lower the energy shared by it and the Earth. So a marble moving toward lower energy would settle in one of the two "troughs" in the track. When you release the marble from points 1 and 2, it's able to get all the way to the bottom trough. When you release the marble from point 3, the best it can do in reaching a low energy is the second trough. Figure 3.5 summarizes this.

What's true for the marble on the track turns out to be a guiding principle in all of science:

Physical systems tend toward configurations that have the lowest energy.

That's a highbrow way of saying that whatever you're looking at, be it a marble on a track or two magnets or a magnet and a ball bearing, tends to be in an arrangement that has the lowest energy possible. Two magnets go to a lower energy by coming together. A ball bearing moves toward a magnet because the magnet and ball bearing have a lower energy when together than when apart.

The word *tend* in the above boldface statement is important. Even though a system tends toward the

Figure 3.5

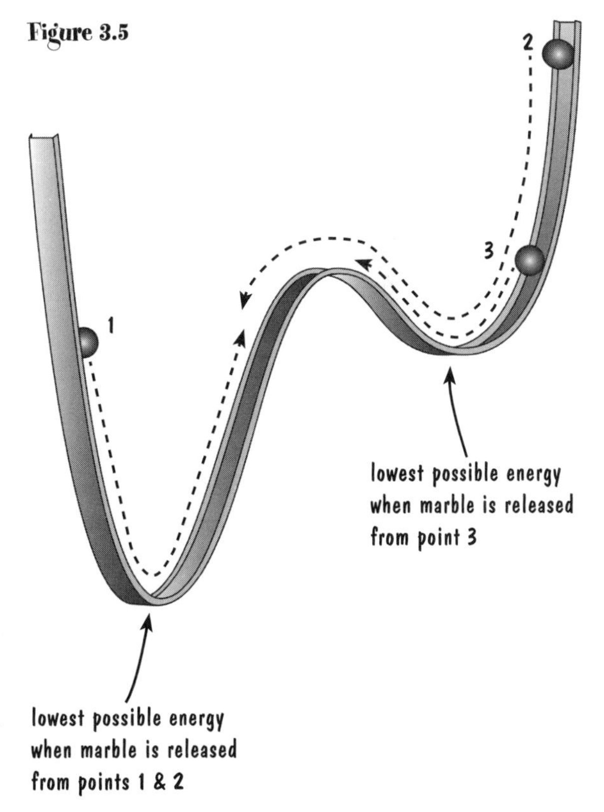

lowest energy, it doesn't always end up with the lowest energy. For example, the lowest energy configuration with the marble and track is for the marble to be at the bottom of the lower trough. When you release the marble from point 3, however, there's no way it can get to the bottom trough. So, it ends up at the lowest energy possible for this situation, which is residing at the bottom of the upper trough. Same thing when you put an index card between the ball bearing and magnet. The ball bearing tends to end up at the lowest possible energy (touching the magnet) but instead ends up at the lowest energy allowed by the situation, which is next to the index card. Check out Figure 3.6.

Figure 3.6

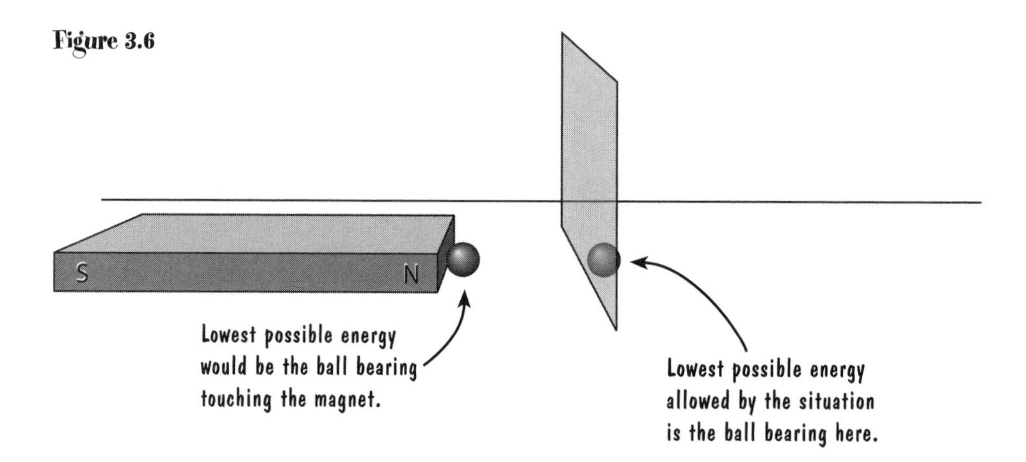

It shouldn't be a big surprise to you that the reason I had you playing with marbles and tracks and magnets is that the concepts also apply to atoms, and also to chemical reactions. The next sections will shed light on this.

More things to do before you read more science stuff

Get about a dozen marbles and a tall drinking glass. Pour the marbles into the glass, so you have something like Figure 3.7.

Figure 3.7

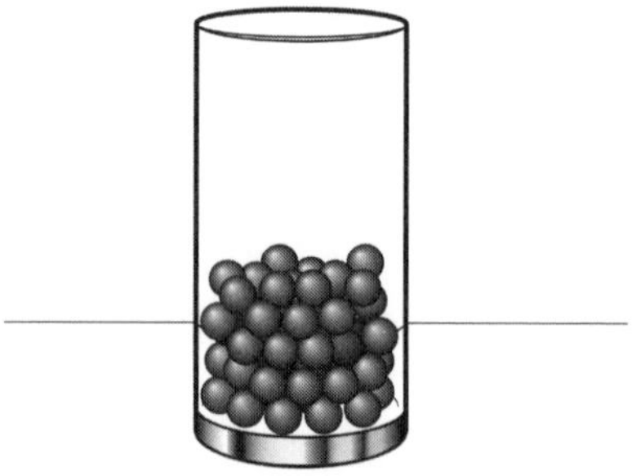

Which of the marbles in the glass have the lowest energy? If you said the ones on the bottom give yourself a gold star. They're closer to the Earth and therefore, when considered together with the Earth, represent the lowest energy. As you move up the glass, the marbles have more and more energy. Next question: Which marbles are the easiest to access? In other words, if you want to interact with a marble or two, which ones would you choose? Score another star if you said the ones at the top of the pile in the glass.

Let's move on to one of my favorite games when I was a kid, namely Stadium Checkers. For all of you who are too young to have played Stadium Checkers (I realize that's most of you!), put the words *stadium checkers* into an internet search engine, and you'll pull up all sorts of pictures of the game, or some version of it. Figure 3.8 provides a drawing.

You begin the game of Stadium Checkers with different-color marbles all positioned at the outermost edge of the "stadium." Each of the concentric rings is free to rotate, and there are slots in the rings that can hold a marble or simply let a marble pass to a lower ring. By rotating the rings, you can move these slots so the marbles move from the outer rings into the inner rings. The object of the game is to get all of *your* marbles into the center before your opponents do the same with their marbles.

Figure 3.8

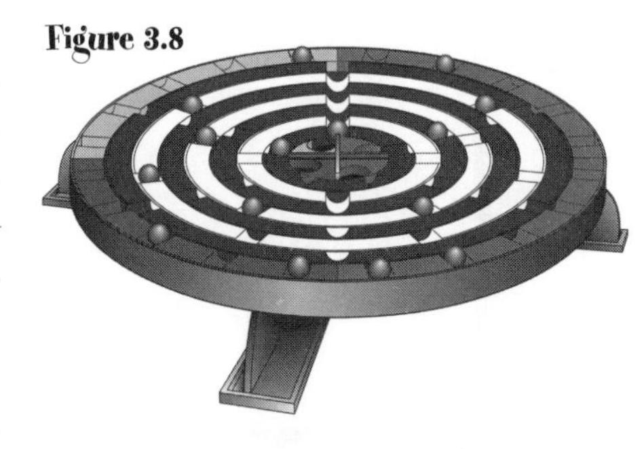

We're not going to play the game, but rather use it to set up a model of atoms. First imagine that the center holes are all blocked, so there is no way to remove a marble from the lowest ring. The best you can do is get a marble *to* the lowest ring. And before moving on, I hope it's clear to you by now that the marbles at the lowest ring have the lowest energy possible, with marbles in successive rings as you move out having higher energies. See Figure 3.9.

Figure 3.9

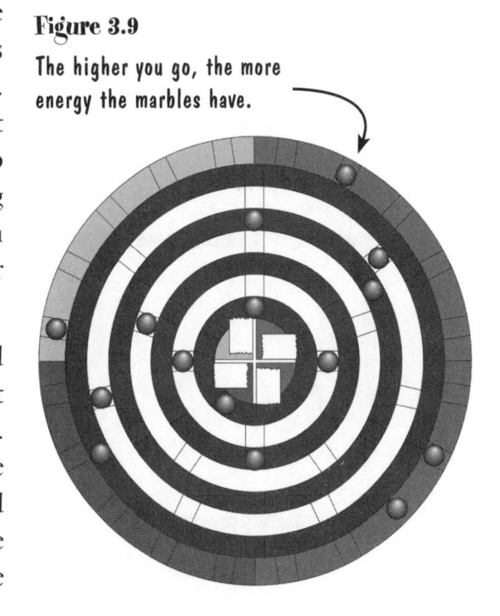

Okay, suppose you have successfully placed four marbles at the lowest ring. Now you want to bring another marble down from the top. Can you put it in the lowest ring? Next suppose you completely fill the second lowest ring, and want to bring another marble down from the top. Can you put this marble in either of the two lowest rings?

While you're pondering those questions, here's a completely different activity. Yes, I'll tie all of this together in the next section! Figure 3.10 (p. 40) shows a floor plan of a house.

Figure 3.10

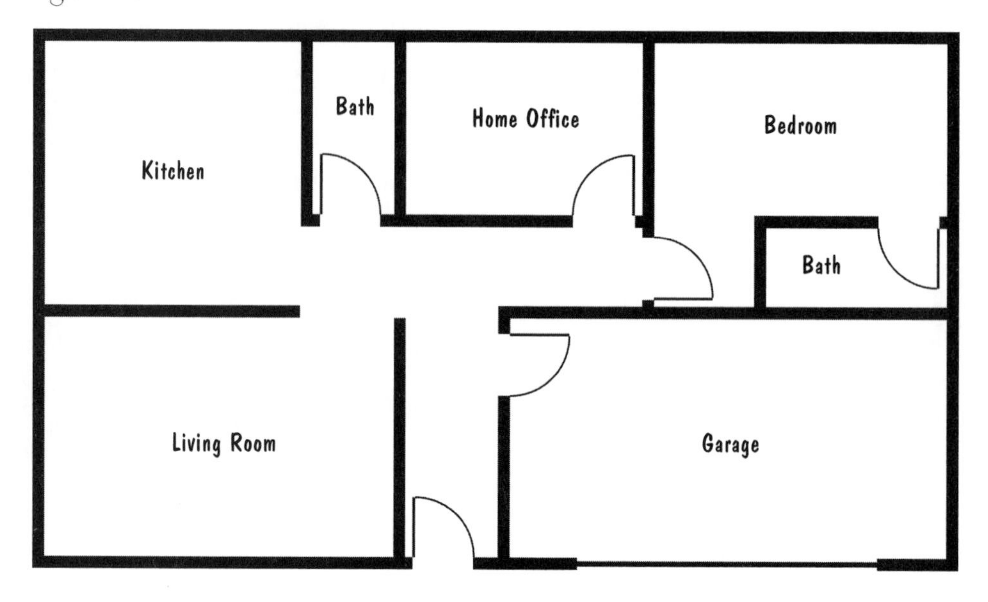

Using shading, I'm going to outline the probability of finding me in various rooms of the house at any time during the Super Bowl. This is shown in Figure 3.11.

Figure 3.11

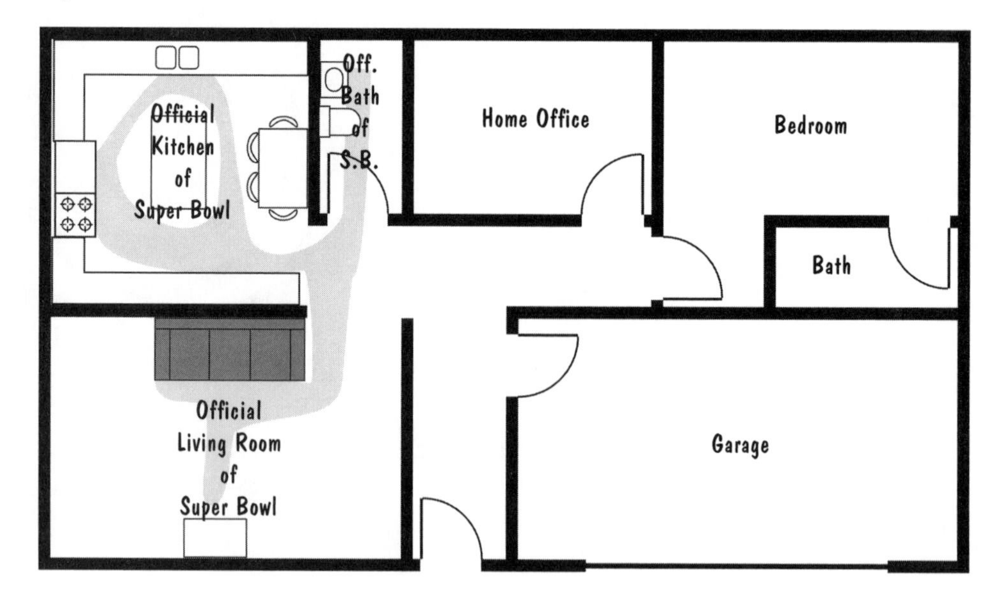

Notice that most of the shading is near the couch in the living room, where I'm parked in front of the television on the sofa. That's because this is the most likely place to find me at any random time. There's also shading, though less of it, in the kitchen (gotta get snacks and beer) and the bathroom. This **probability distribution** shows where Bill is most likely to be during the Super Bowl, and where Bill is totally unlikely to be during the Super Bowl. Note the lack of shading in my home office.

You're turn. Make two copies of Figure 3.10. Use shading to show a probability distribution for the following situations:

- Where you are likely to find someone who is preparing snacks and other goodies for a party of 100 people.
- Where you are likely to find someone who is sick with the flu.

More science stuff

Before we deal with the probability distributions, we can start with the fact that the marbles you played with are a lot like electrons. In Stadium Checkers, you can think of the center of the stadium as the nucleus of the atom, and the marbles as electrons. We already discussed the fact that electrons, being confined within an atom, are only allowed to have certain energies. The different rings of Stadium Checkers correspond to different energies. The analogy goes further. Just as each ring in Stadium Checkers can hold only so many marbles, each energy level in an atom can accommodate only so many electrons. For example, the lowest energy level in any atom can accommodate only two electrons. The second lowest energy state also can accommodate only two electrons. The next highest energy level can accommodate up to six electrons. Once an energy level is filled with electrons, any added electrons must reside in higher energy levels. This is in fact how chemists view the difference between atoms. All atoms have the same basic Stadium Checkers structure. To create a new atom, simply add an electron to an outer ring of the stadium, add a proton to the nucleus so the atom is neutral, and add a neutron or two to the nucleus.

You might ask why a given energy level can only handle so many electrons. The answer lies in the mathematics that describe how things like electrons in atoms behave, and that mathematics is known as **quantum mechanics**. No, I'm not going to teach you quantum mechanics, but I can give you an idea of how it operates. Actually, you already know how it operates with respect to energy. Whenever you confine an oscillation, such as standing waves on a rope, only certain frequencies and their corresponding energies show up as the preferred ones. So we say that the energies appear as discrete values, or *quanta,* of energy. We say the energy is *quantized.* On the scale of electrons, energy isn't the only thing that's quantized. Other physical quantities—things known as "orbital angular momentum" and "spin angular momentum" among other things—are quantized, and can only have certain values. It is this quantization of physical quantities, along with another extremely important principle of quantum mechanics,[3] that limits how many electrons can reside at any particular energy level and also helps define the values of the allowed energy levels. Associated with each quantized quantity are **quantum numbers**, which serve to completely identify the particular state of electrons in atoms. The bottom line for us in this book is that electrons in an atom behave like marbles in a Stadium Checkers game with the bottom holes plugged.

You can represent the location of electrons in an atom with an energy level diagram, such as the one in Figure 3.12 (p. 43). This represents an atom of oxygen, which has eight protons in the nucleus and eight electrons around the nucleus.

A couple of questions. *Why does an atom of oxygen have the same number of protons as electrons?* Electrons have a negative charge and protons have a positive charge. Because things in the universe tend toward being electrically neutral, your garden variety oxygen atom, or any other kind of atom, has the same number of protons and electrons. *Is there anything else in the nucleus of the oxygen atom besides protons?* Yep. You might recall that nuclei contain neutral particles called neutrons in addition to protons.

In Figure 3.12, the energy increases as you go upward. Three energy levels are shown, and they have the labels 1*s*, 2*s*, and 2*p*. I'll explain what the numbers and letters mean later (they represent the aforementioned quantum numbers). For now, note that the two lowest energy levels, 1*s* and 2*s*, each have only two electrons. With two electrons each, these energy levels are completely filled—you couldn't add another electron to these levels if you wanted to.[4] The 2*p* level, on the other hand, is able to hold six electrons. Because oxygen has only eight electrons total, that means that the 2*p* energy level is only partially filled.

Think back to the marbles in the drinking glass. It was easy to interact with the marbles near the top of the pile, but difficult to interact with the marbles near the bottom of the pile. Our oxygen atom is similar. In everyday chemical reactions, those inner electrons in the two lower levels are pretty much untouchable, but those four electrons in the highest energy level are open for business.

[3] The name of this principle is the Pauli Exclusion Principle, which states that no two electrons in an atom can possess identical quantum states. When you study quantum mechanics a bit, you learn that the Pauli Exclusion Principle makes perfect sense; if two electrons had the exact same quantum states, they would in fact *be* the same electron. Call it diversity in physics, if you will.

[4] The reason that *s* energy levels can only hold two electrons each again lies in the details of quantum mechanics. I'm afraid that you'll just have to take such results on faith at this point.

Figure 3.12

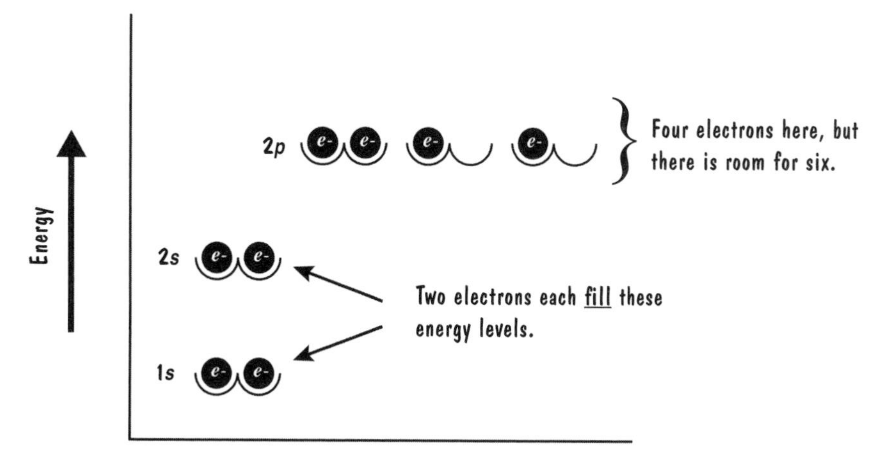

Because there are a couple of slots open in the highest (2*p*) energy level, it's not too far-fetched to imagine electrons from other atoms jumping into those empty slots, or something like that. How these high energy level electrons interact between atoms turns out, in fact, to be the basis for explaining a whole bunch of chemical interactions.

Before explaining those interactions, we can refine our picture of an atom even more by considering those probability distributions you created. My attempts at the two tasks I assigned you are shown in Figures 3.13 and 3.14.

Figure 3.13

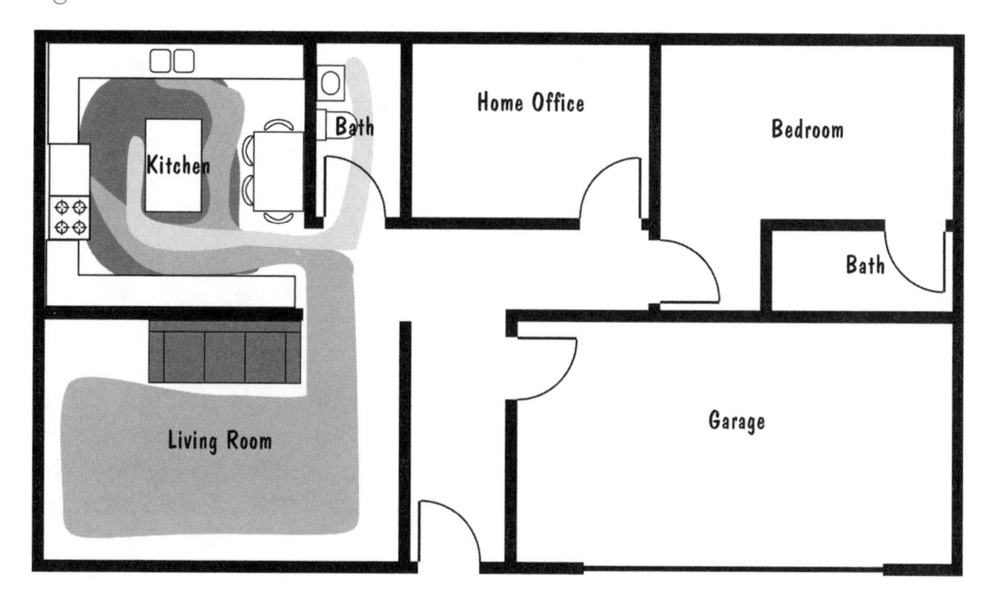

Probability distribution for a person fixing snacks for 100 people.

Figure 3.14

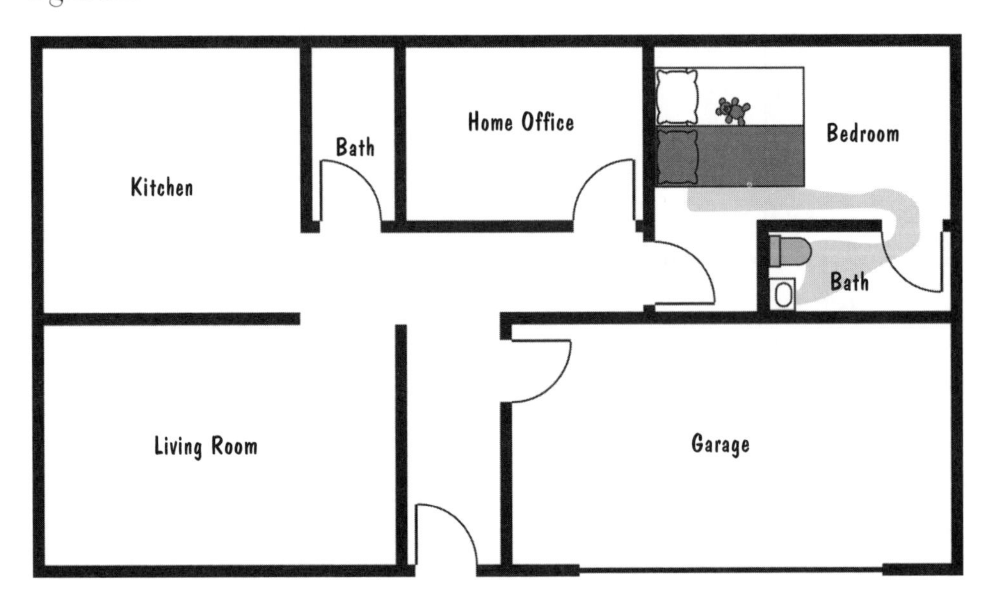

Probability distribution for a person sick with the flu.

If you're preparing snacks for 100 people, it would seem to me that you would spend a lot of time in the kitchen, so that's where I put the most shading. Of course you would also spend a bit of time in other areas of the house, so the shading isn't confined to the kitchen. As for being sick with the flu, personally I spend *all* my time alternating between the bed and the bathroom, so the shading is confined to those places.

What does a floor plan have to do with electrons in atoms? Actually, quite a bit. In quantum mechanics, we represent electrons, and also all sorts of other particles, as bundles of waves. Yes that is strange, but it works. All you need to know about an electron in an atom is represented by something called a *wave function*. By considering energy and forces due to all the other things the electron interacts with, such as protons in the nucleus and other electrons, you can pry information out of the electron's wave function. One of the most useful things to know is where the electron happens to be at any given moment. Unfortunately, that information is unavailable to us. What we *can* determine, using the wave function squared,[5] is a probability distribution for where the electron is. This probability distribution is almost exactly like the probability distributions in our floor plans, with the exception that we can never know for certain where an electron is. You can peek

[5] The quantity you use is technically different from simply squaring the wave function. It's something called the wave function multiplied by its complex conjugate. That distinction isn't important for our purposes, and for that matter, all you need to focus on is the result—the probability distribution.

inside a house and know exactly where a person is, but it doesn't work that way with electrons. All you have is a probability of an electron being in any given spot at any given moment.

Let's consider our oxygen atom. The lowest energy level is 1*s*. The 1*s* energy level has a probability distribution that is a sphere, and it looks something like Figure 3.15.[6]

Figure 3.15

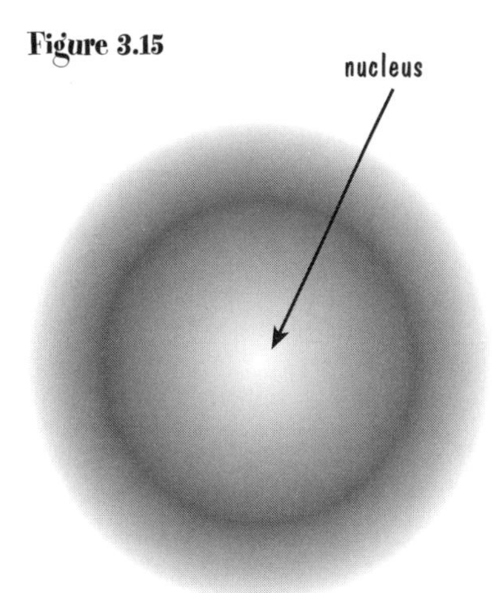

The 1s energy level probability distribution

The nucleus is at the center of this sphere, and the probability distribution has the heaviest shading at a certain distance away from the nucleus. This means that an electron in the 1*s* energy level is most likely to be at this certain distance away from the nucleus, but is equally likely to be anywhere along the sphere at this distance. See Figure 3.16.

> Before moving on, let's stop and take stock of things. We have electrons that are confined to an atom. We found out earlier that *oscillations* confined to a certain space could only have certain allowed energies. We also know that the electrons we're talking about—the ones confined in atoms—are not oscillating in the traditional sense. Through the magic of quantum mechanics, though, we can represent the *electron itself* as a combination of waves, with a corresponding wave function. Confining this wave function to an atom tells us what energy levels are allowed and also gives us probabilities (the probability distribution) for where the electron might be at any given time. We can map out that probability distribution, and for the lowest energy in an atom, the distribution is spherical.

Figure 3.16

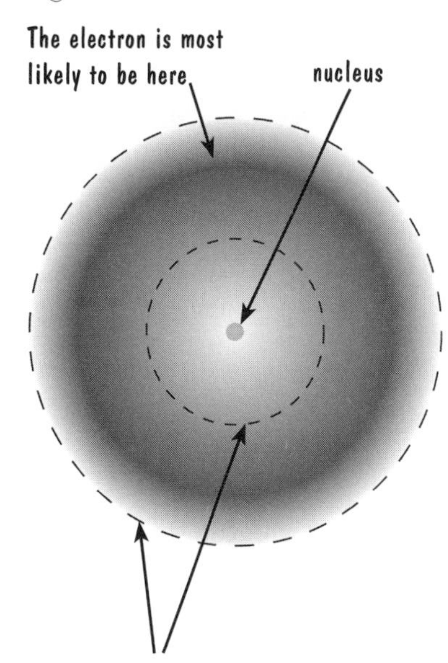

The next lowest energy level in an oxygen atom is the 2*s* level.[7] Like the 1*s* level, it has a spherical shape

[6] The probability distribution for a 1*s* electron actually extends out an infinite distance from the nucleus. The probabilities become extremely small extremely fast, though, so it's conventional to stop the shading at some point. That's a good thing, because we don't have an infinitely large page on which to represent the shading extending to infinity!

[7] Recall that these numbers and letters represent the quantum numbers that define electron energy states. A more thorough explanation for this a couple of pages from now.

and can hold up to two electrons. The 2*s* level is farther out from the nucleus than the 1*s* level, so the 1*s* sphere is inside of the 2*s* sphere. The next energy level up, according to Figure 3.12, is the 2*p* level. When you apply quantum mechanics to this energy level, you end up with a probability distribution that looks sort of like a dumbbell. There are three possible orientations of this dumbbell shape, as shown in Figure 3.17. The different orientations are shown separately. The flat "sheets" in Figure 3.17 are there to clarify the orientation of the dumbbell-shaped figures and are not part of the probability distributions.

Figure 3.17

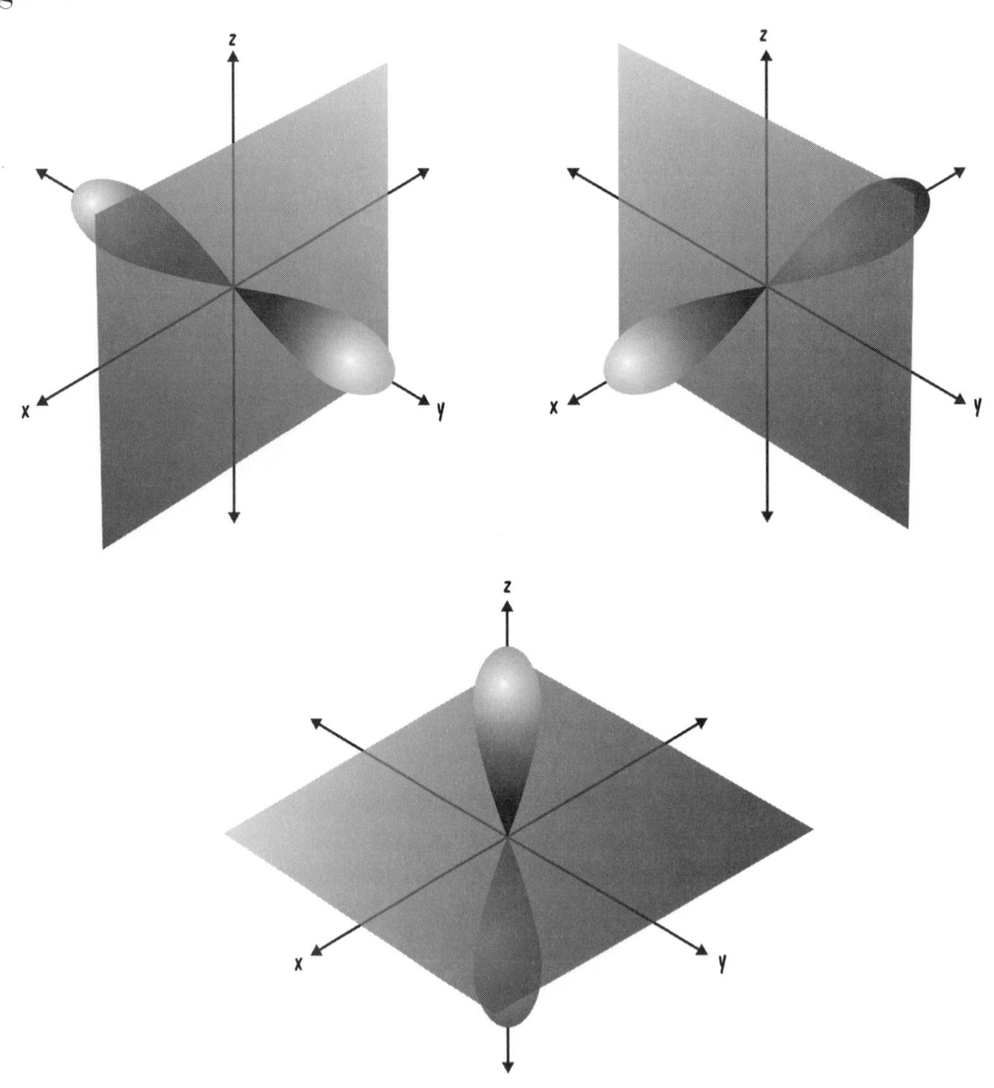

These probability distributions for electrons of different energies are quite useful in determining what atoms will do. The orientations of the *p* levels help determine just how different atoms combine together, for example.

To complete the picture, I need to let you know that atoms contain electron energy levels other than the ones occupied by electrons. This is like a Stadium Checkers game in which you don't have enough marbles to fill all the slots. The slots exist even when they don't hold marbles. Similarly, electron energy levels in atoms exist even if they're not occupied by electrons. A more complete energy level diagram for oxygen is shown in Figure 3.18.

Figure 3.18

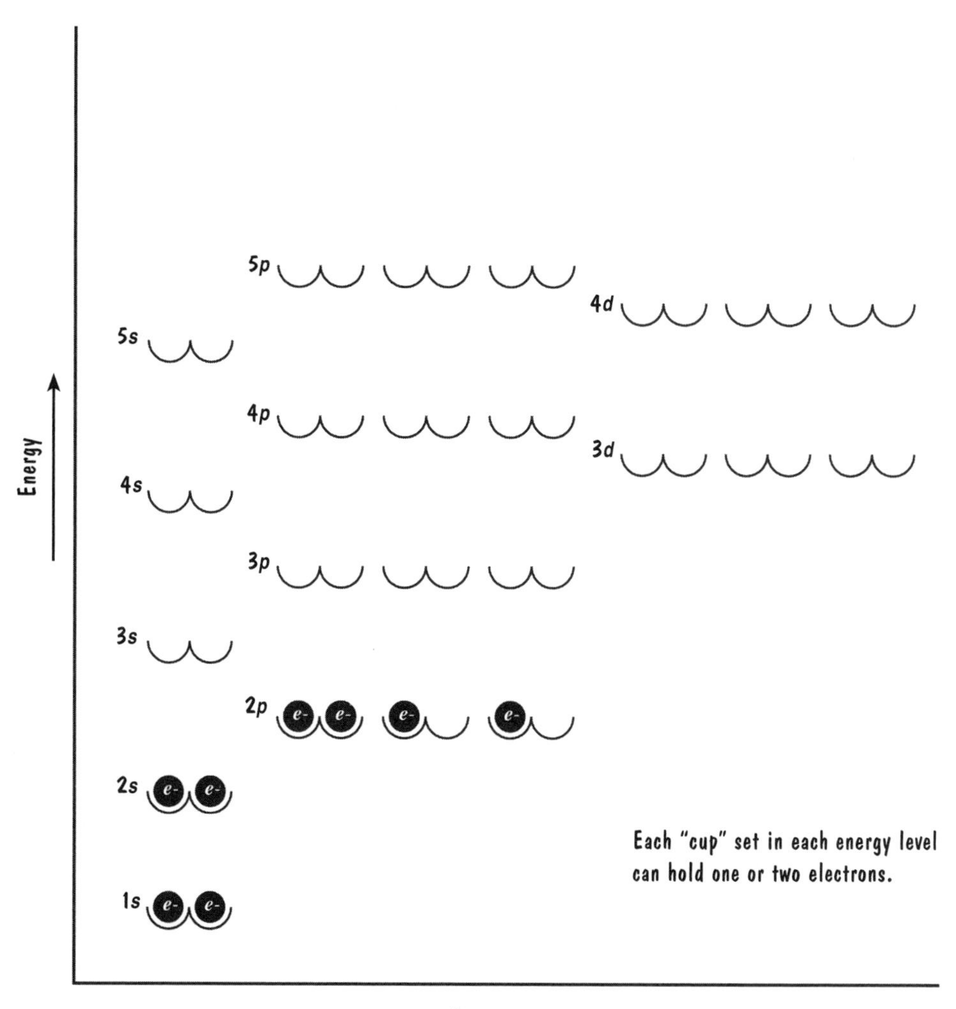

A more complete energy level diagram

Notice that the numbers keep on going up, and the letters change. Different letters correspond to different shapes of probability distributions, and obviously higher numbers mean higher energy. With all these energy levels available, it is entirely possible that electrons can hang out at the higher energy levels. As we learned in the first part of this chapter, though, systems tend toward the lowest possible energy states. Therefore, it's natural for electrons that find themselves at the higher energy levels to jump down to lower energy levels. That often results in the production of light, as we discussed in Chapter 2.

I promised to tell you what the numbers and letters in the energy levels represent, so I guess I better do that. The numbers represent the overall energy level that results because energy is quantized. The 1, 2, 3, 4, etc. are known as **principal quantum numbers**. I told you that other things besides energy are quantized in an atom. One is a quantity known as orbital angular momentum. Without going into what exactly that quantity might be, the small letters *s, p, d,* and *f* represent different quantizations of orbital angular momentum. Okay, so now you know what the numbers and letters stand for, even if the quantities they stand for are not clear.

To end this section, I want to emphasize one major point, which is that our picture of what an atom looks like has progressed quite a bit. We know that there are protons and neutrons in a concentrated nucleus, and we know that there are electrons in various energy levels around that nucleus. Although we cannot say for sure what those electrons are *doing,* we do have drawings that show the probability distribution of electrons of different energies. We also have the concept of filled and partially filled energy levels. Electrons in higher energy levels, especially those in partially filled energy levels, are poised to interact with electrons from other atoms.

Even more things to do before you read even more science stuff

At last we get to the Periodic Table, which is shown in Figure 3.19 (p. 49).

Let's run through what all the numbers and such mean on this table. The letters stand for different elements—H for hydrogen, He for helium, O for oxygen, Be for beryllium, Ag for silver, and on and on and on.[8] Our

Topic: Periodic Table
Go to: *www.scilinks.org*
Code: CB014

[8] You might be troubled by the fact that element symbols sometimes don't seem to represent the element names, such as Ag for silver and W for tungsten. This is because contributions to the positions of elements on the Periodic Table have come from all over the world. The original German name for tungsten, for example, is Wolfram. Hence, W represents tungsten in the table.

Figure 3.19

Periodic Table of Elements

1 H Hydrogen 1.00794																	2 He Helium 4.00260
3 Li Lithium 6.941	4 Be Berylium 9.01218											5 B Boron 10.81	6 C Carbon 12.011	7 N Nitrogen 14.0067	8 O Oxygen 15.9994	9 F Flourine 18.998403	10 Ne Neon 20.179
11 Na Sodium 22.98977	12 Mg Magnesium 24.305											13 Al Aluminum 26.98154	14 Si Silicon 28.0855	15 P Phosphorus 30.97376	16 S Sulfur 32.06	17 Cl Chlorine 35.453	18 Ar Argon 39.948
19 K Potassium 39.0983	20 Ca Calcium 40.08	21 Sc Scandium 44.9559	22 Ti Titanium 47.88	23 V Vanadium 50.9415	24 Cr Chromium 51.996	25 Mn Manganese 54.9380	26 Fe Iron 55.847	27 Co Cobalt 58.9332	28 Ni Nickel 58.69	29 Cu Copper 63.546	30 Zn Zinc 65.38	31 Ga Gallium 69.72	32 Ge Germanium 72.59	33 As Arsenic 74.9216	34 Se Selenium 78.96	35 Br Bromine 79.904	36 Kr Krypton 83.80
37 Rb Rubidium 85.467	38 Sr Strontium 87.62	39 Y Yttrium 88.9059	40 Zr Zirconium 91.22	41 Nb Niobium 92.9064	42 Mo Molybdenum 95.94	43 Tc Technetium (98)	44 Ru Ruthenium 101.07	45 Rh Rhodium 102.9055	46 Pd Palladium 106.42	47 Ag Silver 107.8682	48 Cd Cadmium 112.41	49 In Indium 114.82	50 Sn Tin 118.69	51 Sb Antimony 121.75	52 Te Tellurium 127.60	53 I Iodine 126.9045	54 Xe Xenon 131.29
55 Cs Cesium 132.9054	56 Ba Barium 137.33	57 La Lanthanum 138.9055	72 Hf Hafnium 178.49	73 Ta Tantalum 180.9479	74 W Tungsten 183.85	75 Re Rhenium 186.207	76 Os Osmium 190.2	77 Ir Iridium 192.2	78 Pt Platinum 195.08	79 Au Gold 196.9665	80 Hg Mercury 200.59	81 Tl Thallium 204.383	82 Pb Lead 207.2	83 Bi Bismuth 208.9804	84 Po Polonium (209)	85 At Astantine (210)	86 Rn Radium (222)
87 Fr Francium (223)	88 Ra Radium (226)	89 Ac Actinium (227)	104 Rf Rutherfordium (261)	105 Db Dubnium (262)	106 Sg Seaborgium (266)	107 Bh Bohrium (264)	108 Hs Hassium (269)	109 Mt Meitnerium (268)	110 Uun Ununnium	111 Uuu Unununium (285)	112 Uub Ununbium (284)	113 Uut Ununtrium (289)	114 Uuq Ununquadium (288)	115 Uup Ununpentium (292)	116 Uuh Ununhexium	117 Uus Ununseptium	118 Uuo Ununoctium

58 Ce Cerium 140.12	59 Pr Praseodymium 140.9077	60 Nd Neodymium 144.24	61 Pm Promethium (145)	62 Sm Samarium 150.36	63 Eu Europium 151.96	64 Gd Gadolinium 157.25	65 Tb Terbium 158.9254	66 Dy Dysprosium 162.50	67 Ho Holmium 164.9304	68 Er Erbium 167.26	69 Tm Thulium 168.9342	70 Yb Ytterbium 173.04	71 Lu Lutetium 174.967
90 Th Thorium 232.0381	91 Pa Protactinium 231.0359	92 U Uranium 238.0289	93 Np Neptunium 237.0482	94 Pu Plutonium (244)	95 Am Americium (243)	96 Cm Curium (247)	97 Bk Berkelium (247)	98 Cf Californium (251)	99 Es Einsteinium (252)	100 Fm Fermium (257)	101 Md Mendelevium (258)	102 No Nobelium (259)	103 Lr Lawrencium (260)

notion of an element today really isn't much different from the one proposed by Empedocles. An element is simply something that is a "pure" substance that cannot be considered to be composed of other substances. In our atomic model, an element is something that is composed of only one kind of atom. For the purpose of figuring out the Periodic Table, it will be easiest if you just think of single atoms rather than collections of atoms. So, when we talk about silver, for example, we'll be thinking about a single silver atom rather than a pure block of silver metal.

Anyway, the main number on each element (1 for hydrogen, 2 for helium, 20 for calcium) is called the **atomic number**. It tells you how many protons are in the atom. The number of protons in an atom is the main distinguishing characteristic of an atom. If you change the number of protons, you have a different atom that corresponds to a different element. For a *neutral* atom, the atomic number also tells you the number of electrons in the atom. We'll find out, though, that you can change the number of electrons in an atom without creating a different atom.

The second number by each atom in the Periodic Table is called the **atomic mass**.[9] Atomic masses are not nice, round numbers, but rather numbers such as 28.086 and 65.37. If you truncate the atomic mass (get rid of all numbers to the right of the decimal point), the number you end up with tells you the total number of protons and neutrons in the atom. The atomic mass is measured in what are called *atomic mass units*. In this system, the mass of a proton is exactly equal to 1 and the mass of a neutron is exactly equal to 1.[10] Since the mass of an electron is really small compared to the mass of a proton and the mass of a neutron, we can ignore the electron's mass.

Now take a look at hydrogen in the Periodic Table. Hydrogen contains 1 proton and 1 electron, with zero neutrons. Why, then, isn't the atomic mass of hydrogen exactly equal to 1? The reason is that in nature, each element can have isotopes. Think back to the marshmallow activity and the explanation of isotopes in Chapter 2. Most of the time, hydrogen has 1 proton and zero neutrons, but hydrogen can also have 1 neutron (this is called deuterium) or 2 neutrons (this is called tritium) and still be hydrogen. The atomic masses of elements are not integers because of the existence of isotopes. Atomic masses are *averages* of the atomic masses that naturally occur in nature, and virtually all atoms can exist in isotope form. It's like figuring out the mass of a marshmallow by averaging in the marshmallows that

[9] Older texts and even recent publications use the term *atomic weight* instead of *atomic mass*. The proper term is *atomic mass,* but for years the convention was to use *atomic weight,* and old habits die hard.

[10] The mass of a proton is different from the mass of a neutron, but the difference is small enough that we don't have to worry about it here.

have ball bearings in them. Thus, the atomic mass of hydrogen is 1.00797 instead of 1.00000, and all other atoms have atomic masses that are not integers.

Now that you know what all the numbers represent, it's time to look at the pattern of elements as displayed in the Periodic Table. The first row contains just hydrogen and helium. The second row starts with lithium and ends with neon, but there's a gap in the middle. There's also a gap in the third row. Historically, people began putting the Periodic Table together based on the properties of atoms. The atoms in each *column* are a lot alike in how they interact or don't interact with other atoms. For example, the atoms in the far right column, known as **noble gases**, don't interact very much with other atoms. The atoms in the first and next-to-last column interact strongly with other atoms.

Topic: Noble Gases
Go to: *www.scilinks.org*
Code: CB015

Your task is to see if you can figure out what governs the overall pattern in the Periodic Table. Here's a hint: Look back at the energy level diagram for oxygen (Figure 3.18). The pattern has something to do with what the electrons in the atom are doing. Also, think Stadium Checkers. Of course I'm going to explain this pattern in the next section, but it might be worth your while to take a stab at what's causing the pattern before I tell you.

Even more science stuff

To decipher the pattern present in the Periodic Table, I told you to pay attention to the electrons in the atom and to pay attention to the energy level diagram for oxygen, which is similar to the energy level diagram for other atoms. To refresh your memory, that energy level diagram is repeated in Figure 3.20.

Recall that as you add electrons to energy levels, they fill the lowest energy levels first. Once lower energy levels are filled (remember that energy levels can hold only so many electrons), electrons fill the higher energy levels, just as in Stadium Checkers. Let's use this information to see what happens as the atomic number (number of protons) increases in the Periodic Table.

In hydrogen, as in other atoms, the lowest energy level is 1*s*. Two electrons can reside in that energy level. When you add a proton and two neutrons to hydrogen, you get helium. You also get another electron, which fills the 1*s* energy level. Time to give you another result from quantum mechanics, which is that atoms with nothing but filled energy levels are extremely stable and represent low-energy situations. There's no reason for that to be an obvious result, but it's the result. If you want to add an electron to a filled energy level, clearly that's a difficult thing to do. It's not clear, however, that removing an electron from an entirely filled energy level would be difficult. Just trust me that it's difficult.

Figure 3.20

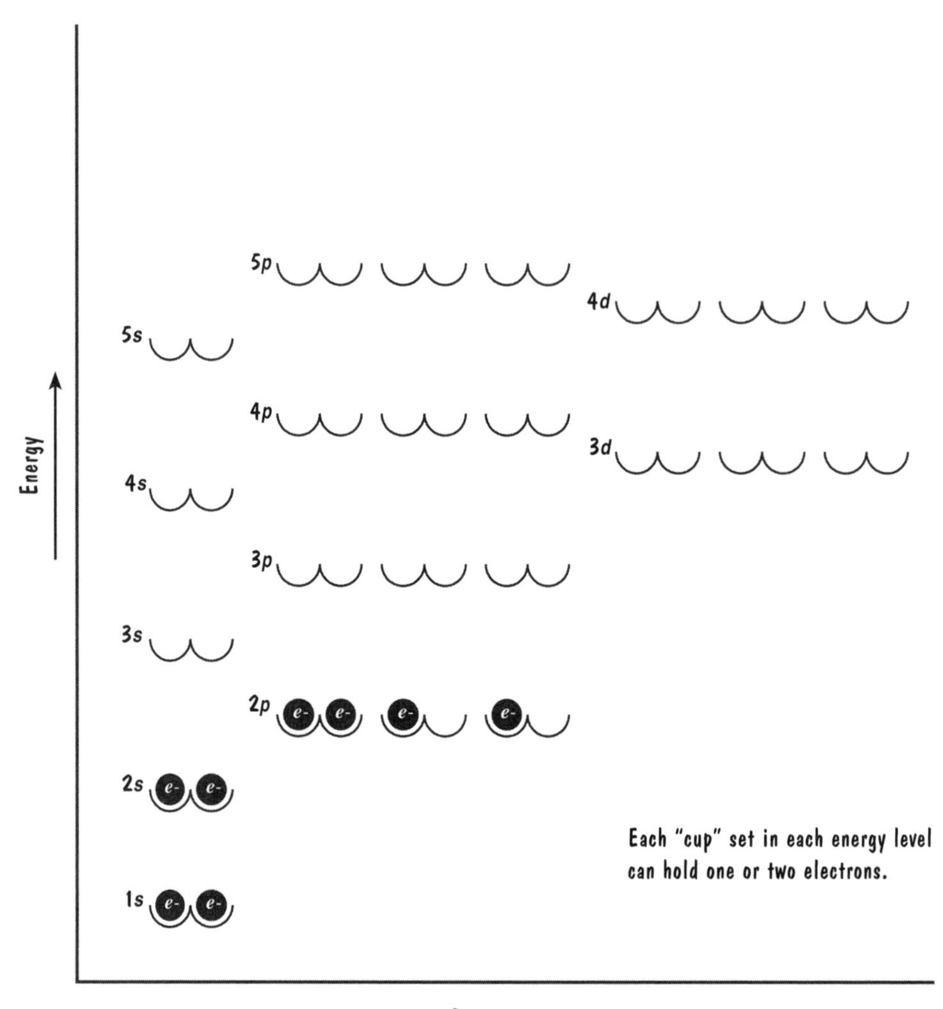

A more complete energy level diagram

SCiLINKS THE WORLD'S A CLICK AWAY
Topic: Inert Gases
Go to: *www.scilinks.org*
Code: CB016

Well, we have helium with a filled 1*s* energy level. Helium is one of those *noble* gases I mentioned earlier. Another name for the noble gases is **inert gases**. The word *inert* means unresponsive or inactive, and that describes helium. Helium just plain doesn't interact much with other atoms. We can explain that non-interaction by the fact that helium has a completely filled energy level. That doesn't explain why you can inhale helium and sound like a Munchkin, but I explain that in another book in this series.

Before continuing, I'm going to introduce new terminology, not to confuse you but because it's commonly used terminology. Scientists refer to the energy levels in atoms as **shells**. Outer shells are higher energy levels and inner shells are lower energy levels. This terminology no doubt dates back to when people thought of electrons as moving in circular orbits around the nucleus, which of course we now know that they don't. Electrons in an outermost shell are known as **valence electrons**. Because electrons in the *s* and *p* energy levels are the most interactive, we only consider *s* and *p* electrons when counting valence electrons. Therefore we would say that hydrogen has one valence electron while helium has two valence electrons.

Back to the Periodic Table. In the second row, we start with lithium. It has a completely filled 1*s* energy level, plus an extra electron that hangs out in the 2*s* energy level. With an open slot in that energy level, lithium readily interacts with other atoms. The next atom in the second row is beryllium, which has a filled 2*s* energy level and thus two valence electrons. Shouldn't be a surprise that beryllium isn't quite as anxious as lithium to interact with other atoms, because it has a filled 2*s* energy level, which is more stable than a half-filled 2*s* energy level.

Next we jump across a gap to the element boron. The reason we jump a gap will be clear later. After you jump the gap, you will notice that there are five elements before you get to the final atom on the right, which is neon. What's happening as you move through those five atoms is that electrons are filling the 2*p* energy level. By the time you get to neon, the 2*p* energy level is filled. Neon has a filled "shell" of eight electrons (recall that the *s* energy level holds up to two electrons and the *p* energy level can hold up to six electrons) and so, like helium, is an inert gas. Filled shells with eight valence electrons don't interact much with other atoms, so neon doesn't take part in many chemical reactions. You might expect that those five atoms before you get to neon—boron, carbon, nitrogen, oxygen, and fluorine—*do* interact with other atoms, and that expectation is correct.

The third row, beginning with sodium and ending with argon, follows the same pattern. It's all about electrons filling the lowest possible energy levels in an atom. The fourth row, which begins with potassium and ends with krypton (birthplace of Superman), all of a sudden has atoms in the gap. Why? Because at this point we begin filling the energy level 3*d*. The 3*d* energy level can hold 10 electrons, which is why there are 10 atoms filling the gap until you get to gallium. At that point, adding protons and their corresponding electrons involves the 4*p* energy level. Because each *p* level holds six electrons, we traverse six atoms in getting to the atom of krypton. Krypton has a filled 1*s* level, a filled 2*s* level, a filled 2*p* level, a filled 3*s* level, a filled 3*p* level, a filled 4*s* level, a filled 3*d* level, and a filled 4*p* level.

I should say something about the two rows at the very bottom that are separated from the rest of the Periodic Table. These represent a progression of elements in which electrons are filling *f* energy levels. If we included them in the rest of the Periodic Table, there would be another huge gap that would stretch the table from side to side. Then we wouldn't be able to fit a readable Periodic Table on a page. It's sort of like how Alaska and Hawaii are portrayed on a map of the United States. They're not shown in their actual locations because if they were, the rest of the map would not be readable on a normal sized page.

Okay, that's enough to describe the general pattern. As you move from left to right across the Periodic Table, you are adding protons and electrons to atoms. As this happens, the electrons, in cooperation with the nuclei of atoms, obey the simple principle that systems naturally tend to be in their lowest energy. See Figure 3.21.

So, as you go from left to right in the Periodic Table, electrons are filling available slots in various energy levels. As one energy level gets filled with electrons,

Figure 3.21

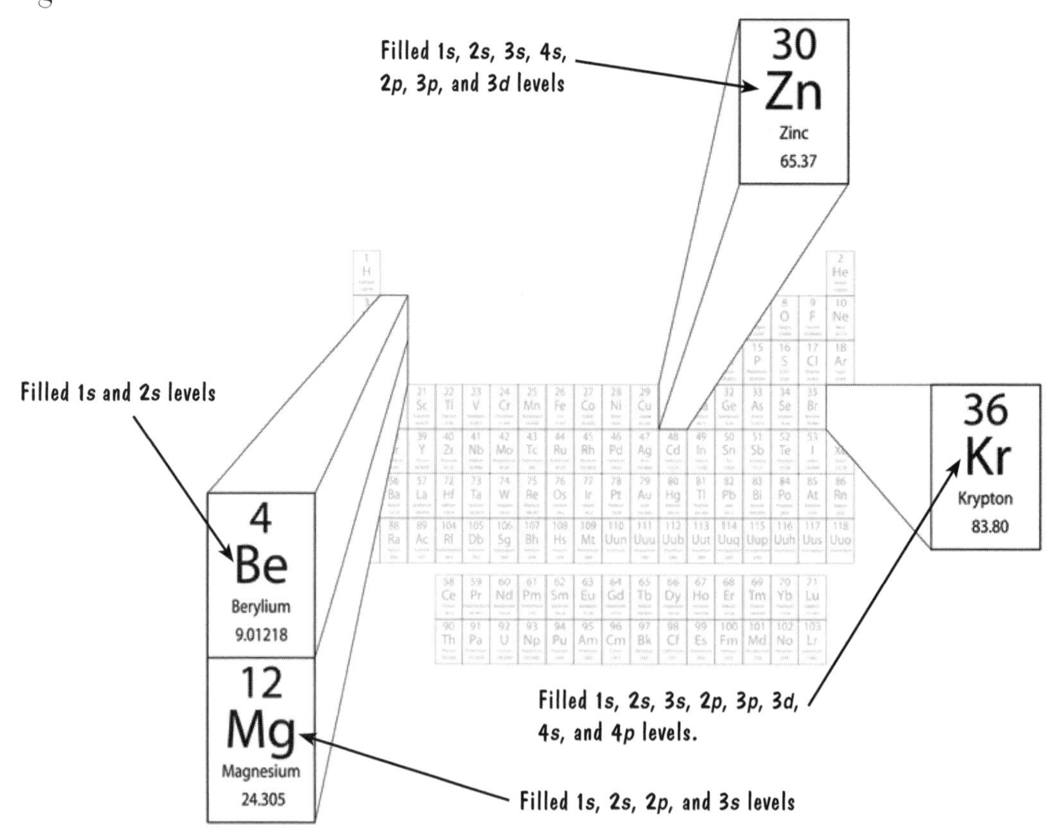

you start on the next energy level up. Once you have this picture, you can figure out lots of chemical reactions. As I said before, electrons are the key to understanding what happens when different atoms get together. I also told you that atoms with unfilled energy levels more readily interact with other atoms.

> Keep in mind that when we speak of moving from one atom to another on the Periodic Table, we're not literally moving from one atom to another. We are simply using the wealth of information in the table to make comparisons regarding the characteristics of different atoms.

Now let's look at columns in the Periodic Table. Each atom in the far left column has a half-filled *s* energy level. Hydrogen has one electron in its 1*s* energy level, and the 1*s* level can hold two electrons. Lithium has one electron in its 2*s* energy level, and the 2*s* level can hold two electrons. Sodium has one electron in its 3*s* energy level, and the 3*s* level can hold two electrons. Potassium has one electron in its 4*s* energy level, and the 4*s* level can hold two electrons. You can continue this pattern all the way down the first column in the Periodic Table. Each of these atoms has an *s* energy level that is half full. Check out Figure 3.22.

Figure 3.22

Each atom in the first column has a half-filled *s* energy level.
Each atom in the last column has nothing but filled energy levels.

1 H Hydrogen 1.00794																	2 He Helium 4.00260
3 Li Lithium 6.941	4 Be Beryllium 9.01218											5 B Boron 10.81	6 C Carbon 12.011	7 N Nitrogen 14.0067	8 O Oxygen 15.9994	9 F Flourine 18.998403	10 Ne Neon 20.179
11 Na Sodium 22.98977	12 Mg Magnesium 24.305											13 Al Aluminum 26.98154	14 Si Silicon 28.0855	15 P Phosphorus 30.97376	16 S Sulfur 32.06	17 Cl Chlorine 35.453	18 Ar Argon 39.948
19 K Potassium 39.0983	20 Ca Calcium 40.08	21 Sc Scandium 44.9559	22 Ti Titanium 47.88	23 V Vanadium 50.9415	24 Cr Chromium 51.996	25 Mn Manganese 54.9380	26 Fe Iron 55.847	27 Co Cobalt 58.9332	28 Ni Nickel 58.69	29 Cu Copper 63.546	30 Zn Zinc 65.38	31 Ga Gallium 69.72	32 Ge Germanium 72.59	33 As Arsenic 74.9216	34 Se Selenium 78.96	35 Br Bromine 79.904	36 Kr Krypton 83.80
37 Rb Rubidium 85.467	38 Sr Strontium 87.62	39 Y Yttrium 88.9059	40 Zr Zirconium 91.22	41 Nb Niobium 92.9064	42 Mo Molybdenum 95.94	43 Tc Technetium (98)	44 Ru Ruthenium 101.07	45 Rh Rhodium 102.9055	46 Pd Palladium 106.42	47 Ag Silver 107.8682	48 Cd Cadmium 112.41	49 In Indium 114.82	50 Sn Tin 118.69	51 Sb Antimony 121.75	52 Te Tellurium 127.60	53 I Iodine 126.9045	54 Xe Xenon 131.29
55 Cs Cesium 132.9054	56 Ba Barium 137.33	57 La Lanthanum 138.9055	72 Hf Hafnium 178.49	73 Ta Tantalum 180.9479	74 W Tungsten 183.85	75 Re Rhenium 186.207	76 Os Osmium 190.2	77 Ir Iridium 192.2	78 Pt Platinum 195.08	79 Au Gold 196.9665	80 Hg Mercury 200.59	81 Tl Thallium 204.383	82 Pb Lead 207.2	83 Bi Bismuth 208.9804	84 Po Polonium (209)	85 At Astantine (210)	86 Rn Radium (222)
87 Fr Francium (223)	88 Ra Radium (226)	89 Ac Actinium (227)	104 Rf Rutherfordium (261)	105 Db Dubnium (262)	106 Sg Seaborgium (266)	107 Bh Bohrium (264)	108 Hs Hassium (269)	109 Mt Meitnerium (268)	110 Uun Ununnium	111 Uuu Unununium (285)	112 Uub Ununbium (284)	113 Uut Ununtrium (289)	114 Uuq Ununquadium (288)	115 Uup Ununpentium (292)	116 Uuh Ununhexium	117 Uus Ununseptium	118 Uuo Ununoctium

58 Ce Cerium 140.12	59 Pr Praseodymium 140.9077	60 Nd Neodymium 144.24	61 Pm Promethium (145)	62 Sm Samarium 150.36	63 Eu Europium 151.96	64 Gd Gadolinium 157.25	65 Tb Terbium 158.9254	66 Dy Dysprosium 162.50	67 Ho Holmium 164.9304	68 Er Erbium 167.26	69 Tm Thulium 168.9342	70 Yb Ytterbium 173.04	71 Lu Lutetium 174.967
90 Th Thorium 232.0381	91 Pa Protactinium 231.0359	92 U Uranium 238.0289	93 Np Neptunium 237.0482	94 Pu Plutonium (244)	95 Am Americium (243)	96 Cm Curium (247)	97 Bk Berkelium (247)	98 Cf Californium (251)	99 Es Einsteinium (252)	100 Fm Fermium (257)	101 Md Mendelevium (258)	102 No Nobelium (259)	103 Lr Lawrencium (260)

Each atom in the first column has a half-filled *s* energy level. Because electrons are the main ingredient in chemical reactions, you might expect that the atoms in the first column behave similarly in chemical reactions, and they do.

Take a look at the far right column of the Periodic Table. Each atom in this column has all of its energy levels filled by electrons. That makes for stable atoms that don't interact much with other atoms. So, all those atoms in the far right column should behave similarly in chemical reactions, right? Right. That's what happens.

This general rule is the same for all columns of the Periodic Table. Magnesium, calcium, and strontium have similar properties. Copper, silver, and gold have similar properties. Fluorine, chlorine, bromine, and iodine have similar properties.

To recap, you add electrons and steadily fill electron energy levels, or shells, as you move from left to right across the Periodic Table. All atoms in a particular column of the Periodic Table have similar properties, because they have the same number of valence electrons (electrons in the outermost *s* and *p* energy levels).

Historically, the Periodic Table came to be as a result of scientists classifying elements according to their properties. Elements in each column behave similarly. The meaning behind this classification becomes clear once we have a picture of how additional electrons fill energy levels, and how the arrangement of those electrons affects what elements do. My approach here has been to develop a current model of atoms, and then use that to explain the organization of atoms and elements in the Periodic Table. Hopefully that will help you do less memorizing and more understanding.

Chapter Summary

- Physical systems, including atoms, tend to reach a configuration that has the lowest possible energy for the situation.
- Electrons in atoms reside in discreet energy levels, which we refer to as electron shells. The mathematics of quantum mechanics dictate that different electron shells can hold up to a specific number of electrons. Once a lower energy level shell is filled, further electrons have to reside in higher energy shells.
- Quantum mechanics further dictates that we cannot know exactly where electrons are, but rather can only calculate probability distributions that tell us the probability of an electron being in a given place. These probability distributions have definite shapes, and these shapes affect how atoms interact with one another.
- Electrons in the outer *s* and *p* shells of an atom are called valence electrons.

- The Periodic Table gives us a great deal of information about the atoms that make up elements. From the table, we can determine how many protons and neutrons are in an atom as well as the configuration of electrons in an atom.
- The atoms in a single column of the Periodic Table tend to have similar properties. This is because atoms in the same column have similar electron configurations.

Applications

1. In Chapter 2, I explained how atoms emit light when electrons jump from a higher energy level to a lower energy level. We can use this, plus the fact that systems tend toward lower energy states, to explain why, for example, objects glow when they get hot enough. When you heat an object, you are adding energy in the form of heat to the atoms of that object. That energy can cause electrons to move up to higher energy levels. Now, because atoms naturally head toward lower energies, the electrons that jumped up to the higher energy levels then spontaneously jump back down to lower energy levels. In the process, they emit light.
2. It's not impossible for "inner shell" electrons (those tucked away at lower energy levels when an atom contains electrons at higher energy levels) to interact with the outside world, but it takes a lot of energy for this to happen. Valence electrons in the outer shell, however, are easy to interact with. One reason has to do with what is known as **electron shielding**. A negatively-charged electron in an outer shell is held in the atom by the electric force between it and the positively-charged nucleus. However, there are often many negatively-charged electrons in lower energy levels between this valence electron and the nucleus. These repel the valence electron, and effectively cancel most of the attractive force between the nucleus and the valence electron. See Figure 3.23.

Figure 3.23

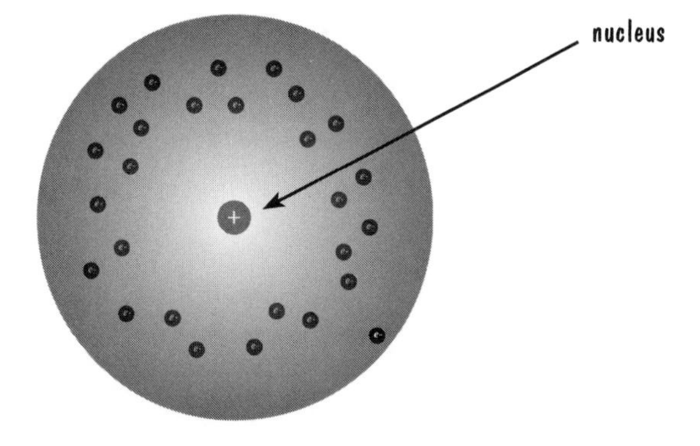

An outer electron is "shielded" from the nucleus by other electrons.

3. Just as *s* and *p* energy levels have definite shapes to their probability distributions, *d* energy levels have a definite shape to their probability distributions. The five possible *d*-level probability distributions are shown in Figure 3.24. These shapes are the result of calculations in quantum mechanics. Note again that the flat sheets are there to clarify the orientation of the probability distributions.

Figure 3.24

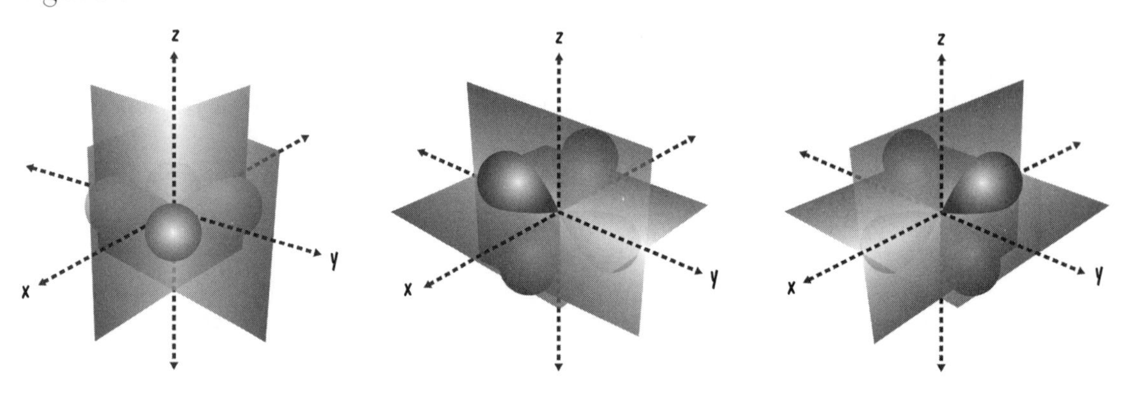

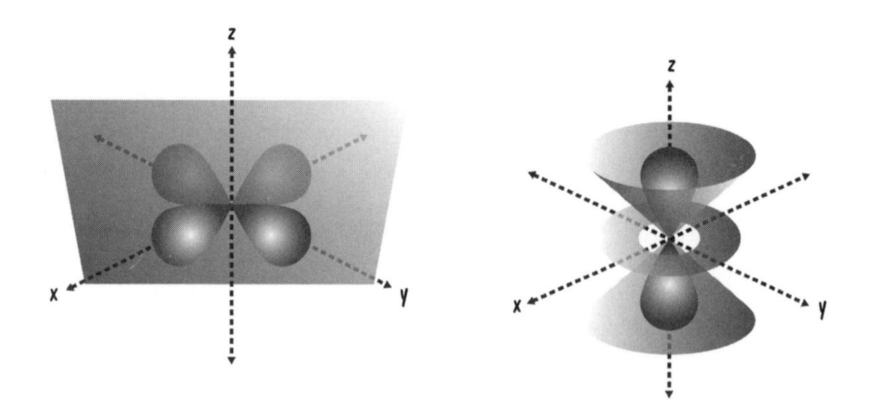

Let's Get Together ... Yeah, Yeah, Yeah

Ah yes, the theme song from one of the strangest movies of all time—*Parent Trap*. The original, that is, with Hayley Mills. I was in love with her back then. So naturally I would choose that phrase to title a chapter on how atoms combine to form molecules. Sometimes atoms of the same kind combine, but most often it's different kinds of atoms. There are three main ways that atoms get together, and we'll discuss all three in this chapter.

Things to do before you read the science stuff

Still have that apparatus you built for determining which materials conduct electricity? Sure you do. If not, head back to Chapter 2 for the instructions on how to put together another one. Once you have this, place the two open ends (the bare wires) into a glass or cup of water. Make sure the two bare wires don't touch each other. Does the water conduct electricity? Nah, because the bulb doesn't light.[1] Next dissolve a whole bunch of salt in the water. Test the saltwater to see if it conducts electricity. Hmmm. What do you suppose is going on? Doesn't the conduction of electricity require the movement of charged things? Did you charge up the water? Did you charge up the salt? No, of course you didn't.

Figure 4.1

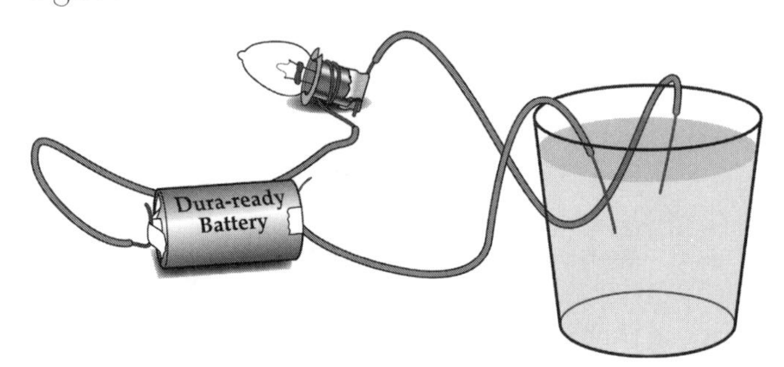

The science stuff

Allow me to explain why saltwater conducts electricity. Salt itself is electrically neutral (equal numbers of positive and negative charges), as is the water. When you dissolve salt in water, however, that changes. Salt is composed of sodium and chlorine atoms, hence the name of sodium chloride for salt. When you dissolve salt in water, much of the salt separates into separate sodium and chlorine atoms, but there's a twist. Because chlorine atoms have a stronger attraction for electrons than sodium atoms, the chlorine atoms end up with an extra electron and the sodium atoms end up minus an electron. Atoms that are charged positively or negatively are called **ions**. So, it's the sodium ions and chlorine ions that conduct electricity in the saltwater solution.

> Note that changing the number of electrons in an atom doesn't change the identity of an atom. The number of protons is the key to identity. Every atom that has 11 protons is a sodium atom, regardless of the number of electrons added or lost.

[1] We're interested in the difference in electrical conduction between normal and salt-filled water. The "normal" water in your area might already contain a lot of dissolved salts so the lightbulb lights before you add salt. If that happens, note the difference in brightness of the bulb with and without salt in the water.

Okay, so what? Ions in saltwater conduct electricity. Here's what. This activity demonstrates that some atoms like electrons more than other atoms.[2] Because the chlorine atoms like electrons more than sodium atoms, they both become ions. The fact that sodium and chlorine atoms have different affinities for electrons explains what happens when many atoms get together. Because we've been talking about sodium and chlorine, let's stick with that example.

If you heat up the metal sodium until it's molten and then expose it to chlorine gas, you end up with salt. (**Don't actually do this!** This is a dangerous activity, only to be done with strict safety precautions.) Let's figure out how that happens. First, sodium is in the far left column of the Periodic Table. That means it has a half-filled 3*s* energy level. If it gains an electron, then it has a filled 3*s* energy level, and that's a good thing. On the other hand, if it loses an electron, sodium has an empty 3*s* energy level and all other energy levels filled, which is a good thing. It turns out that that latter situation is energetically more favorable—lower energy states overall for each ion. In the reaction with chlorine, sodium loses an electron and becomes positively charged. It's still sodium, because the number of protons doesn't change, and the number of protons is what determines what atom you have.

Chlorine is the other atom involved. To get to an energetically favorable situation, chlorine can gain one electron to get a completely filled outer shell of electrons, or it can lose five electrons to get to the same situation. Clearly, the best bet for chlorine is to gain just one electron, and that's what it does when interacting with sodium. So, what happens is that the sodium atom steals an electron from the chlorine atom. Now the two ions are attracted to each other because they have opposite charges. This happens with a whole bunch of sodium and chlorine atoms, and you end up with the molecule known as sodium chloride, or table salt. The atoms bond together through electric forces, and this is known as an **ionic bond**. Figure 4.2 (p. 62) illustrates the process.

Topic: Ionic Bonding
Go to: *www.scilinks.org*
Code: CB017

Let's imagine bringing two other atoms together, such as two oxygen atoms. These two oxygen atoms are identical, so it would be silly to expect one of them to "like" electrons more than the other. Even so, each oxygen atom has two empty slots in its outer shell, and it would be nice for them to be able to fill the outer shell. Instead of forming an ionic bond, which really isn't possible since there's no tendency for an electron or two to jump from one atom to the other, the

[2] Some people get really upset about anthropomorphic statements in science. I don't. When I speak of atoms liking or not liking electrons, I'm just referring to the relative strength of the electric attraction between the atoms and electrons.

Figure 4.2

Cl–

Chlorine steals an electron from sodium and they attract because of electric forces.

Na+

Many of these ionic bonds form, creating the molecule that is table salt.

oxygen atoms do something different. They *share* two of their valence electrons. By sharing, each atom can almost have a filled outer shell. I say *almost* because it's not the same as each atom having two extra electrons all to themselves. The sharing of these two electrons closely resembles filled shells though, so this situation is energetically favorable. When two atoms share one or more electrons, it's called a **covalent bond** ("co" as in together and "valent" as in valence electrons). The atoms are bonded because they're in an energetically favorable situation. It's like a marble being at the bottom of a trough, as in the beginning of Chapter 3. This energy trough might not be the lowest one available to oxygen atoms in the universe, but it's still a trough. It would take an input of energy to break such a bond. See Figure 4.3.

Topic: Covalent Bond
Go to: *www.scilinks.org*
Code: CB018

Figure 4.3

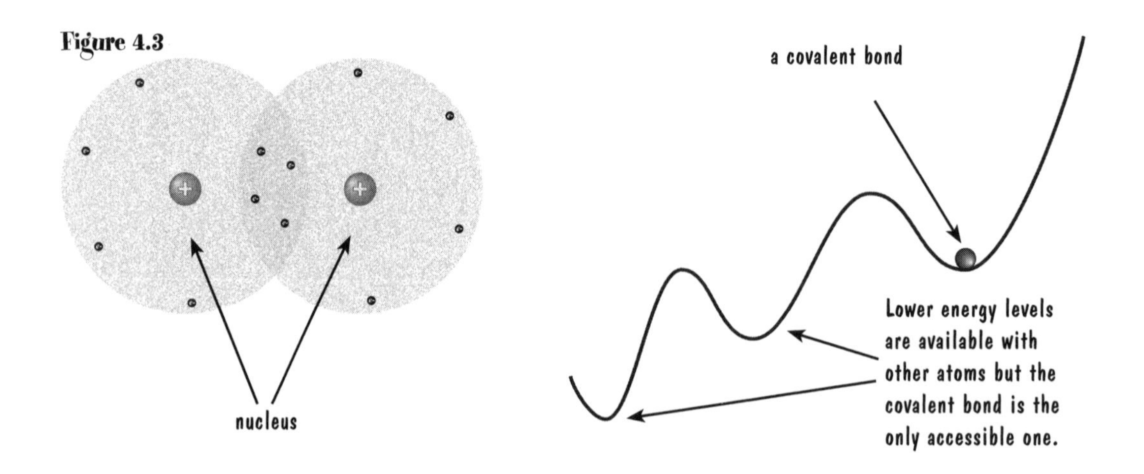

Let me summarize before moving on. Sometimes atoms get together and an electron or two jumps from one atom to another. Then the two atoms have opposite charges and feel an attractive force. This kind of bond is called an ionic bond. It turns out, by the way, that ionic bonds form between metals and nonmetals. More on that a bit later. The second kind of bond we know about is a covalent bond, where atoms share electrons. These kinds of bonds occur between two nonmetals.

More things to do before you read more science stuff

Blow up a balloon and tie it off. Find yourself a faucet and start running a stream of water. Rub the balloon on your hair or a sweater or a favorite pet, and then bring the balloon near the stream of water. You should get a deflection of the water, as shown in Figure 4.4.

Figure 4.4

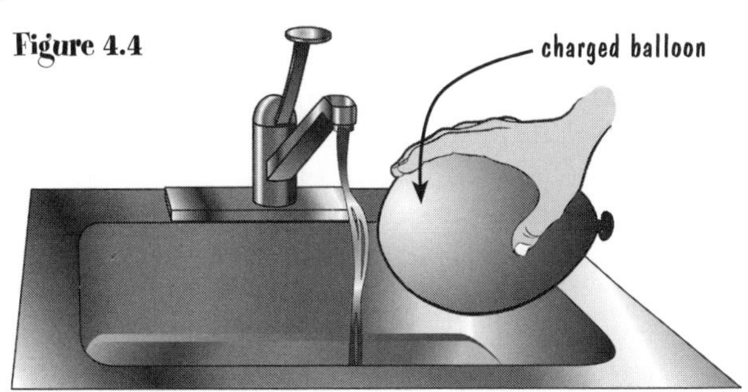

Water is composed of two hydrogen atoms and one oxygen atom. Because both of these are nonmetals, they're connected by covalent bonds. Each hydrogen atom shares two electrons with the oxygen atom, which is illustrated in Figure 4.5.

Figure 4.5

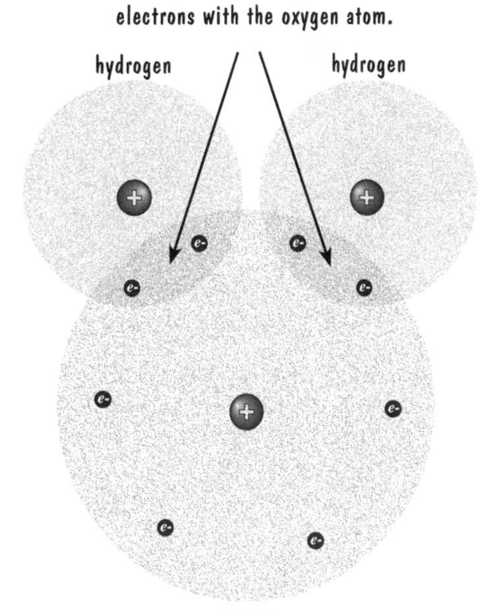

Now here's a question. The entire water molecule contains an equal number of protons and electrons, and so is neutral. How come a charged balloon attracts water molecules? Think about that a bit before moving on to the next section.

More science stuff

Even though an oxygen atom and two hydrogen atoms form covalent bonds, oxygen still likes electrons more than hydrogen. So, the atoms share, but not equally. The shared electrons spend more of their time near the oxygen atom than they do the hydrogen atom. If we think back to the idea of probability distributions, this means that the probability distribution for

Figure 4.6

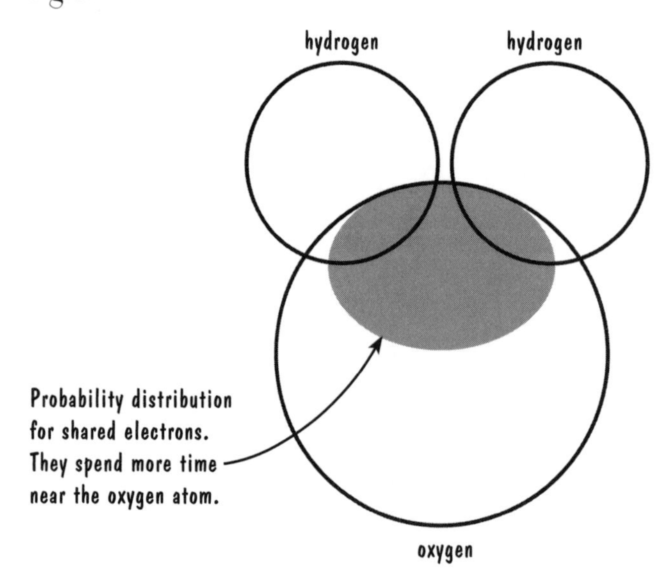

the shared electrons is shaded more near the oxygen than near the hydrogen. See figure 4.6.

Now we can understand why water is attracted to a charged balloon. Because of the uneven distribution of shared electrons, the water molecule is *polar,* meaning one end is positive and the other negative.[3] When you bring a charged object (the balloon) near the water molecules, the positive end is attracted to the extra electrons on the balloon, and the negative end is repelled. Because electric forces get weaker with distance, the attractive force is stronger than the repulsive force, and the water molecules move toward the balloon. Check out Figure 4.7 (p. 65).

The covalent bond between hydrogen and oxygen is called a **polar covalent bond**. The bond is covalent because the atoms are sharing electrons, but it's also polar because the shared electrons spend more time near the oxygen atom than near the hydrogen atoms. Compare this with our oxygen molecule, consisting of two oxygen atoms. This is also a covalent bond, but the shared electrons do not spend more or less time near either of the atoms. This bond is called a **nonpolar covalent bond**.[4]

Topic: Electronegativity
Go to: *www.scilinks.org*
Code: CB019

Well, how do you predict what kind of bond will form between two kinds of atoms? Fortunately for us, chemists have calculated a special number that describes the extent to which atoms like electrons when they are bonded with other atoms. It's called **electronegativity**. Figure 4.8 (p. 66) shows the Periodic Table with the electronegativities of the atoms labeled.[5]

[3] Also critical in water being a polar molecule is the fact that it is not symmetrical. If the oxygen atom were for some reason surrounded by evenly spaced hydrogen atoms (which it isn't), then the molecule would not be polar.

[4] A nonpolar covalent bond is also sometimes called a "pure" covalent bond.

[5] There are several different electronegativity scales. The one shown in Figure 4.8 represents the Pauling scale. The Pauling scale is less effective than other scales at predicting exactly how strong a bond will form in a molecule, but it is the most commonly used one. Hence, we'll use that scale and remember that this scale will give us general trends and not necessarily exact results. More on this in the Applications section.

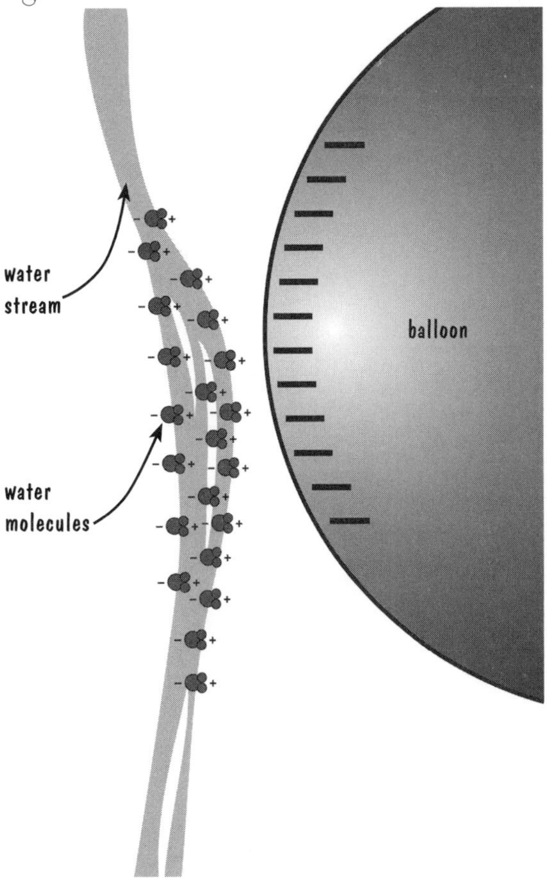

The larger the number associated with electronegativity, the greater affinity the atom has for extra electrons when bonded with another atom. So fluorine, with an electronegativity of 4.0, likes electrons a lot when bonded with other atoms. Cesium, with an electronegativity of 0.7, is not all that fond of electrons when bonded with other atoms. There are general trends to notice and understand with electronegativities. First, the far right column of the Periodic Table isn't even included in Figure 4.8. That's because the noble gases in that column are notoriously unreactive with other atoms since they have completed *s* and *p* energy levels. Noble gases *can* form bonds with other atoms, but it's a rare occurrence, so we're safe in ignoring those situations.

Figure 4.7

water stream

water molecules

balloon

The next thing to notice is that atoms in the far right column in Figure 4.8 (the next to last column in the periodic table) have among the highest electronegativities, while the atoms in the far left column have among the lowest electronegativities. This should make sense to you, because the atoms in the far right column of Figure 4.8 need only one electron to complete their outer shell. That means they will have a strong affinity for one more electron when forming a bond with another atom. Of course, this is related to the fact that a filled electron shell represents a low energy, and atoms tend toward the lowest energy state. The atoms in the far left column need lose only one electron to revert to a situation where they have a completed outer shell, so they readily lose an electron.

In general, electronegativities increase as you move from left to right across the Periodic Table, which is due to what I discussed in the previous paragraph. There's another pattern, though. As you go *down* the Periodic Table, electronegativities tend to decrease. We can explain that one, too. In Chapter 3, I discussed *electron shielding,* in which outer electrons are shielded from the nucleus by all the electrons that are in inner shells. The farther you go down the columns in the periodic table, the more electrons you have participating in this electron shielding. Plus, the atoms farther down in a column are larger (more electrons), so the outermost

Figure 4.8

Electronegativities of the Elements

1 H Hydrogen 2.1																
3 Li Lithium 1.0	4 Be Berylium 1.5											5 B Boron 2.0	6 C Carbon 2.5	7 N Nitrogen 3.0	8 O Oxygen 3.5	9 F Flourine 4.0
11 Na Sodium 0.9	12 Mg Magnesium 1.2											13 Al Aluminum 1.5	14 Si Silicon 1.8	15 P Phosphorus 2.1	16 S Sulfur 2.5	17 Cl Chlorine 3.0
19 K Potassium 0.8	20 Ca Calcium 1.0	21 Sc Scandium 1.3	22 Ti Titanium 1.5	23 V Vanadium 1.6	24 Cr Chromium 1.6	25 Mn Manganese 1.5	26 Fe Iron 1.8	27 Co Cobalt 1.9	28 Ni Nickel 1.9	29 Cu Copper 1.9	30 Zn Zinc 1.6	31 Ga Gallium 1.6	32 Ge Germanium 1.8	33 As Arsenic 2.0	34 Se Selenium 2.4	35 Br Bromine 2.8
37 Rb Rubidium 0.8	38 Sr Strontium 1.0	39 Y Yttrium 1.2	40 Zr Zirconium 1.4	41 Nb Niobium 1.6	42 Mo Molybdenum 1.8	43 Tc Technetium 1.9	44 Ru Ruthenium 2.2	45 Rh Rhodium 2.2	46 Pd Palladium 2.2	47 Ag Silver 1.9	48 Cd Cadmium 1.7	49 In Indium 1.7	50 Sn Tin 1.8	51 Sb Antimony 1.9	52 Te Tellurium 2.1	53 I Iodine 2.5
55 Cs Cesium 0.7	56 Ba Barium 0.9	57 La Lanthanum 1.1	72 Hf Hafnium 1.3	73 Ta Tantalum 1.5	74 W Tungsten 1.7	75 Re Rhenium 1.9	76 Os Osmium 2.2	77 Ir Iridium 2.2	78 Pt Platinum 2.2	79 Au Gold 2.4	80 Hg Mercury 1.9	81 Tl Thallium 1.8	82 Pb Lead 1.9	83 Bi Bismuth 1.9	84 Po Polonium 2.0	85 At Astantine 2.2
87 Fr Francium 0.7	88 Ra Radium 0.9	89 Ac Actinium 1.1														

electrons are a relatively long way from the nucleus. Electric attraction gets weaker the farther away you are. So, the outer electrons aren't held as strongly the farther you move down a column. Hence the decrease in electronegativities.

One last thing regarding electronegativity. You can use these numbers to predict what kind of bond will form between two atoms. When two atoms share electrons and these two atoms are similar in electronegativity, the bond is nonpolar covalent. This tends to occur between two similar elements, such as two nonmetals. They share electrons about equally between their nuclei. When two atoms with a greater difference in electronegativity get together, they form a polar covalent bond. If the difference in electronegativities is really large, then one electron might completely jump over to the other atom, and an ionic bond forms.[6] There are general ranges in electronegativity you can use to predict the kind of bond formed. If two nonmetals get together and the difference in their electronegativities is less than 0.2, they will form a *nonpolar covalent bond.* If the electronegativity difference is from 0.3 to 1.4, they will form a *polar covalent bond.*[7] If a metal and a nonmetal get together and the electronegativity difference is 1.5 or greater, they will form an *ionic bond.*

Even more things to do before you read even more science stuff

This is a short section. See if you can figure out what kinds of bonds I've ignored in the previous section. We have nonmetals bonding to nonmetals in covalent bonds and we have metals bonding to nonmetals in ionic bonds. Don't tax your brain *too* much in figuring out the bonds that are left out.

Even more science stuff

This is an equally short section. What I left out were metals bonding to other metals. We know this happens because pure metal elements exist in reasonably large pieces, and something must be holding all those atoms together. Metal atoms don't just share electrons with one or two other metal atoms—they share all of their valence electrons with all the other metal atoms in a piece of metal. This

[6] You will find many chemists and chemistry books claiming that ionic bonds are not technically chemical bonds because there are no shared electrons. Instead, these bonds are simply electric attractions. I'm going with the traditional view, though, so I consider ionic bonds to be chemical bonds.

[7] As mentioned in a previous footnote, you also have to consider the symmetry of the situation. An electronegativity difference in this range will not result in a polar bond if the molecule is symmetrical.

has been described as a **sea of electrons**, where electrons roam about more or less freely from one metal atom to another. It's sort of an electron commune; the valence electrons of each atom belong to the entire group of nuclei.

The "sea of electrons" picture explains why metals have the properties they do. Metals conduct electricity and heat easily. Given that electricity, and sometimes heat, is due to the motion of charged particles like electrons, it makes sense that free electrons in metals would lead to the conduction of electricity and heat. Although I'm not going to go into detail here, the fact that metals have lots of free electrons also explains why metals tend to reflect light and other electromagnetic waves, such as radio waves, well. The light hits the metal and interacts with the free electrons, which readily reradiate the light back out as reflected light.

And even more things to do before you read even more science stuff

Aluminum and chlorine can combine to form a salt—aluminum chloride. This is typical of what happens when metals and nonmetals get together. Take a look at the Periodic Table, and notice that aluminum has a filled 3*s* energy level and one electron in the 3*p* energy level. In other words, it has three valence electrons. If it's going to form an ionic bond with chlorine, it will most likely give up all of its valence electrons to chlorine. Trouble is, chlorine needs only one electron to fill its outer shell. Your task: Figure out how many chlorine atoms combine with one aluminum atom when aluminum chloride forms.

Another task. If you leave various metals open to the elements, they often develop a cloudy layer on the surface. With the metal iron, we call that layer **rust**. With aluminum, we call it a layer of **oxidized aluminum**. The reason for this name is that the aluminum on the surface combines with oxygen, to form a compound called **aluminum oxide**. Your task is to figure out how many aluminum atoms combine with how many oxygen atoms. This is harder than your previous task, because oxygen has six valence electrons. It only needs two more electrons to complete its outer shell, but aluminum has three valence electrons to give. How can this work?

And even more science stuff

Maybe you figured out how to combine those atoms, and maybe not. Well, here are the answers in case you didn't. First the aluminum chloride. The aluminum atom has three valence electrons to give, but the chlorine atom can only accept one electron. The solution is to use three chlorine atoms for each aluminum atom. That way, each chlorine atom takes one of the three valence electrons. We write the resulting molecule with subscripts that indicate how many of each atom are involved in the molecule. Because there is one aluminum atom and there are three chlorine atoms, aluminum chloride is written as $AlCl_3$. The subscript of 1 on the aluminum is implied.

You've undoubtedly seen this kind of symbol before, if only with water molecules. Because each water molecule contains two hydrogens and one oxygen, water is written as H_2O. Similarly, hydrogen gas, which consists of two hydrogen atoms, is written as H_2. Oxygen gas is written as O_2.

Now we move on to aluminum oxide. We know that one aluminum atom has three valence electrons to give, and oxygen can accept two valence electrons. We can make this work if we adjust the number of atoms. If we use two aluminum atoms, that gives us a total of six valence electrons. Because each oxygen atom can accept two of these, then three oxygen atoms will accept all six of those valence electrons offered up by the two aluminum atoms. So, we end up with two aluminum atoms and three oxygen atoms, and the molecular formula Al_2O_3.

The point of these exercises is to show that most molecules are not simple combinations of one atom of one element and one atom of another element. Things get complicated.

Chapter Summary

- Different atoms have different affinities for the electrons in their outer shells.
- Electronegativity is a number assigned to each atom that describes the affinity an atom has for its valence electrons.
- The different affinities different atoms have for electrons lead to three main types of bonds—nonpolar covalent bonds, polar covalent bonds, and ionic bonds.
- Covalent bonds involve the sharing of electrons between atoms.
- Ionic bonds result from an electron jumping from one atom to another. This leads to an electrostatic attraction that is the bond.
- You can use a knowledge of the number of valence electrons in atoms and a knowledge of the electronegativities of atoms to determine not just what kinds of bonds will form between atoms, but also how many of each atom will be involved in that bond.
- Atoms in a metal bond together through the sharing of all their valence electrons among all the atoms of the metal. This is often referred to as a sea of electrons.

Applications

1. Helium is lighter than air, which is why air pushes helium upward, making helium-filled balloons float. The Periodic Table easily explains why helium is lighter than air. A helium molecule is monatomic, which means that it is composed of just one atom. The atomic mass of a helium atom is about 4 (ignoring isotopes), so that is also its molecular mass. Now let's see how much mass air has. Air is composed primarily of nitrogen and oxygen. The molecular mass of a nitrogen molecule (N_2) is 28 because each nitrogen atom has an atomic mass of 14. The molecular mass of an oxygen molecule (O_2) is 32 because each oxygen atom has an atomic mass of 16. Clearly, then, helium has a whole lot less mass, and therefore weighs less than the major components of air. Providing there are equal numbers of helium and air molecules in a given space, the heavier and denser air will push the lighter and less dense helium gas upward.

2. If you've ever messed around with carbon dioxide gas, you know that carbon dioxide gas sinks in air. A simple way to demonstrate this is to mix baking soda and vinegar in a cup. You did this in Chapter 1. This produces carbon dioxide gas. Light a candle. If you tilt the cup as if you're pouring the contents out, without actually pouring out any of the liquid, and pour over the candle, the candle will go out. See Figure 4.9.

   The reason the candle goes out is that the "poured" carbon dioxide deprives the flame of oxygen, which it needs to burn. Why does carbon dioxide sink in air? Its molecular formula is CO_2, meaning it has one carbon atom and two oxygen atoms. The atomic mass of carbon is 12. The atomic mass of oxygen is 16, and there are two of these atoms in the carbon dioxide molecule. Therefore, the total molecular mass of carbon dioxide is 44. Compare this with the molecular mass of oxygen and nitrogen (first application), and you can see why carbon dioxide sinks in air.

Figure 4.9

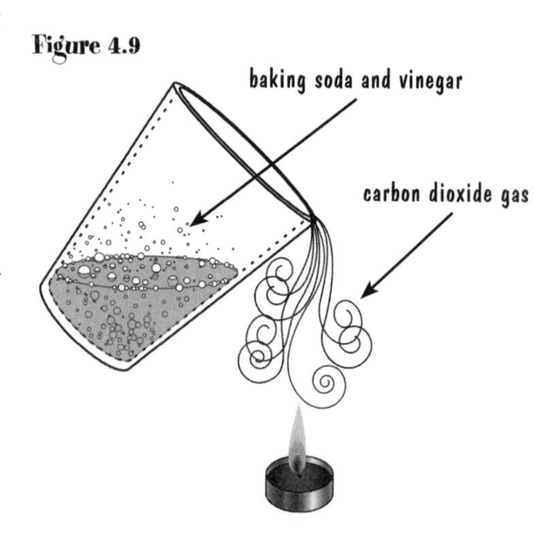

"Pouring" carbon dioxide gas over a candle.

3. All we've talked about so far are two different kinds of atoms getting together and forming a molecule. We're going to jump ahead just a bit, though, and talk about bringing two different molecules together and allowing them to react with each other. The two molecules are lithium iodide (LiI) and cesium

fluoride (CsF). These are both salts, because they're combinations of metals and nonmetals. Mix these salts together and you end up with two new molecules, namely lithium fluoride (LiF) and cesium iodide (CsI). Why does this switch happen? The answer lies partially in electronegativity values and partially in the size of the atoms. Here are the electronegativity values: Fluorine—4.0, lithium—1.0, iodine—2.5, and cesium—0.7. Recall that the strength of an ionic bond can be determined by the difference in electronegativity values between two atoms. The electronegativity difference for lithium iodide is 1.5, and the electronegativity difference for cesium fluoride is 3.3. Now look at the products of the reaction I described above. The electronegativity difference for lithium fluoride is 3.0, and the electronegativity difference for cesium iodide is 1.8. The atoms switched places, but you might be wondering why. We gain a stronger bond in one case, going from 1.5 to 1.8, but we lose in going from 3.3 to 3.0. Why is this reaction energetically favorable? To see why, you have to recall (see an earlier footnote) that the Pauling scale of electronegativity provides guidelines for the strength of bonds, but not complete accuracy. The *size* of the atoms must be considered. Lithium is a small atom compared to cesium, because cesium contains so many more electrons. When lithium loses an electron, as it does in forming an ionic bond, there aren't a lot of electrons left to shield the nucleus from other atoms. Also, the resulting lithium ion is much smaller than the cesium ion, so the positive and negative ions can be closer together when lithium is involved in the ionic bond. So, that positive charge of the lithium nucleus makes a strong bond with negatively charged ions. Cesium, on the other hand, is rather large and has lots of electrons that shield the nucleus from negative ions. So, even though cesium loses an electron more easily than lithium, it forms weaker bonds with negative ions than does lithium. In other words, the Pauling scale of electronegativity doesn't tell the whole story. To get the entire story, you need to include the radius, or size, of the involved atoms. See Figure 4.10.[8]

Figure 4.10

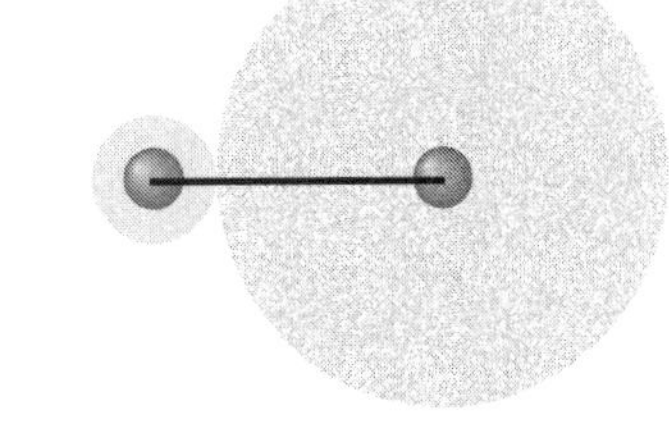

When a small atom bonds with a large atom, the electric attraction is strong because the charges are close together.

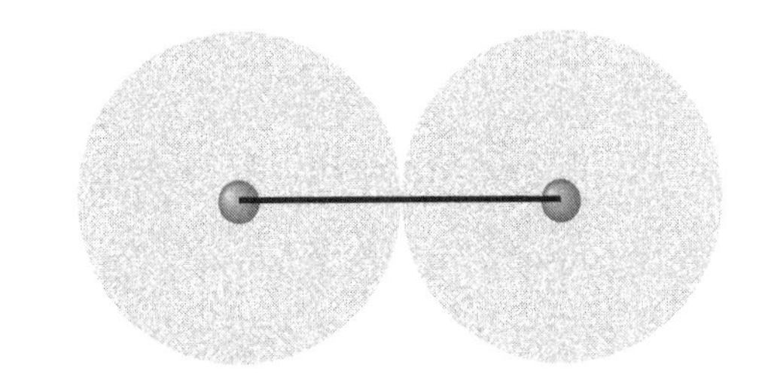

When a large atom bonds with a large atom, the electric attraction is weak because the charges are farther apart.

[8] There is an electronegativity scale, called the Allred-Rochow scale, which does take the radii of the atoms into account. Chemists rely on this scale more than on the Pauling scale. As I said, though, the Pauling scale is most often used in textbooks, so that's the one I've used here. Better for you to understand why the scale included in an introductory chemistry text doesn't tell the whole story than to use a scale here that won't translate to your use of a text.

Therefore, when you mix the original salts, the stronger bond of lithium fluoride will naturally form more readily than other bonds. The cesium and iodine left over will go ahead and form their own ionic bond, though it is a weak bond. Chemists often refer to hard ionic bonds and soft ionic bonds. The hard bonds have a large electronegativity difference plus an advantage in the relative size department and the soft bonds have a smaller electronegativity difference and a disadvantage in the relative size department. In chemical reactions, the hard bonds always dominate over the soft bonds.

4. In the beginning of this chapter, you learned that putting table salt, or sodium chloride, into water causes the sodium and chloride ions to separate. The reason they do this is that water is a polar molecule, with one side positive and the other negative (remember, it's a polar covalent bond). The positive sides of the water molecules surround the negative side of the salt molecule (the chloride ion) and the negative sides of water molecules surround the positive side of the salt molecule (the sodium ion). In this way, the salt *dissolves* in water. Figure 4.11 (p. 73) illustrates what happens. The polar nature of water molecules makes water a useful solvent for many types of molecules.

Figure 4.11

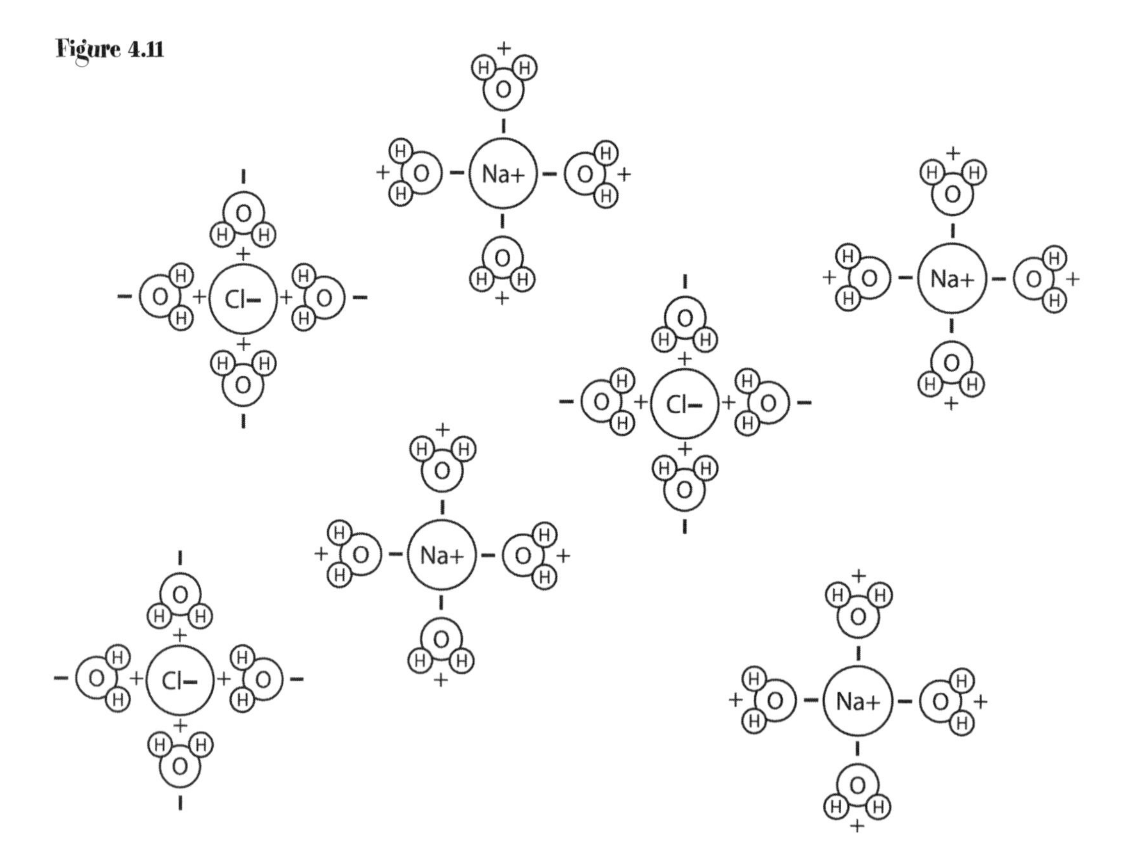

5. Perhaps you recall from the chemistry course you took in high school something called the Lewis dot formula, or just the electron-dot representation. In this representation, atoms are shown with little dots all around them, and there are rules for what happens to the dots when atoms combine. My wife recalls this in particular as one of those things she had to memorize in chemistry but had *zero* understanding of what she was doing. Well, the electron-dot representation is pretty simple if viewed in terms of what valence electrons are doing. For example, Figure 4.12 shows two Hs, which represent two hydrogen atoms, each with one dot near the H. The dot represents the single valence electron possessed by a hydrogen atom. When you combine the two atoms, they form a covalent bond, sharing their valence electrons. This covalent bond is represented by both dots being between the Hs.

Figure 4.12

Two separate hydrogen atoms, each with one valence electron.

Two hydrogens sharing their valence electrons.

All other atoms are represented with up to eight dots around the letters, representing up to eight valence electrons. A few electron-dot representations for single atoms and for combinations of atoms are shown in Figure 4.13.

Figure 4.13

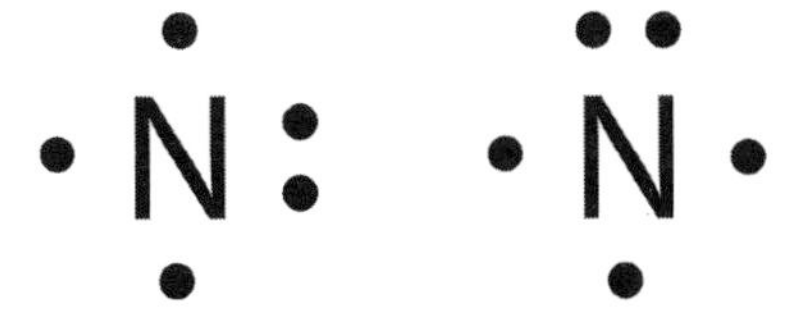

Two separate nitrogen atoms, each with three empty slots in its outer shell.

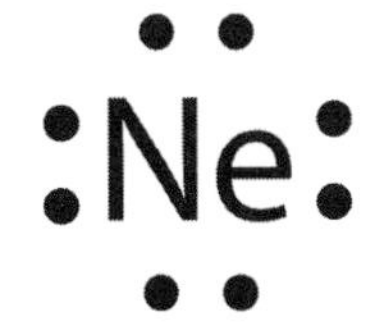

The noble gas neon with a completely filled outer shell.

:N:::N:

A nitrogen molecule with two nitrogen atoms sharing six electrons.

So, those dots shouldn't be mysterious. They simply represent valence electrons, and as we now know, valence electrons and their interactions are at the heart of most chemical reactions.

Balancing Act

We now know enough about atoms and how they combine to investigate a number of different chemical reactions. We'll even be able to represent those reactions with symbols that tell us how many of each kind of atom we begin with and how many of each kind we end up with. This is a powerful tool, and finally gets us to the point of improving upon Empedocles' elements of earth, fire, air, and water. Before we begin, though, I want to impress upon you how much time and effort went into developing the models we're going to use. In this chapter, I'm going to give you molecular formulas for all kinds of substances. Those formulas don't just come out of thin air, but rather represent tons of experiments, replications of experiments, and intense thought as to how the atoms in the Periodic Table come together and how we can decide what kinds of atoms and molecules are present in a situation.

Things to do before you read the science stuff

The first activity I'm going to have you do is a repeat of an activity in the *Stop Faking It!* book on air, water, and weather. I'm repeating it because in addition to relating back to the burning of a match (Chapter 1 of this book), it will serve as an example of just how much information we can get from a chemical equation that represents a chemical reaction. Get a votive candle, a small pan of water, and a glass that's large enough to fit over the candle but small enough in diameter that it will fit inside the pan. Put about a centimeter's depth of water in the pan and place the candle in the water. Clearly we don't want so much water that it covers the candle (see Figure 5.1).

Figure 5.1

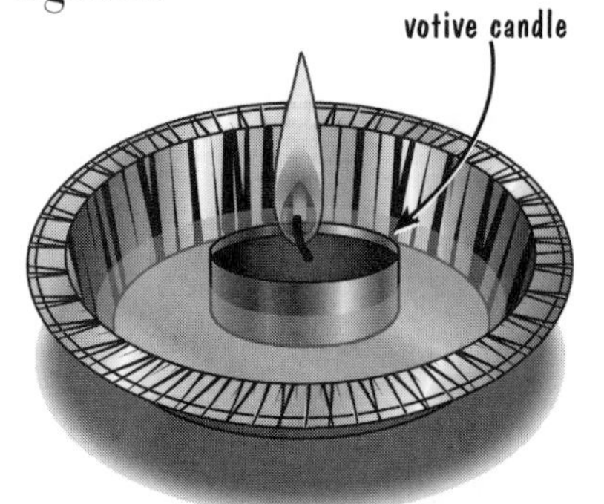

Figure 5.2

Now light the candle and place the glass upside down in the pan over the candle, as in Figure 5.2. As the candle goes out, you should notice that the water rises up inside the glass. Also notice that tiny water droplets form on the inside of the glass above the main water level.

The science stuff

Here's the part where I pull a few molecular formulas out of thin air. It's beyond the scope of this book to go into exactly how we know that the chemicals involved have the formulas given, or for that matter how we know that the chemicals I mention are the ones that are involved. I promise to go into such matters in the second chemistry book in the *Stop Faking It!* series.

Before explaining the burning of a candle, I'm going to explain a slightly simpler burning process—that of burning the gas methane, which is the gas used in gas stoves, gas furnaces, and Bunsen burners. Whenever any substance burns, it combines with oxygen molecules and produces carbon dioxide and water. The molecular formula for oxygen is O_2 (you already knew that), the molecular formula for methane is CH_4 (one carbon atom and four hydrogen atoms—maybe you didn't know that), the molecular formula for water is H_2O (two hydrogens and an oxygen—you knew that one too), and the molecular formula for carbon dioxide is CO_2 (one carbon and two oxygens—you might have known that). I told you the chemical reaction is that molecules of oxygen combine with molecules of methane to produce molecules of carbon dioxide and molecules of water. We can represent that as follows:

(oxygen) + (methane) → (carbon dioxide) + (water)

This is a crude form of what is known as a **chemical equation**. It doesn't have an equals sign, so you might be wondering why it's an equation. I'll explain that in a bit. First, you need to know a bit of terminology. The things on the left of this relationship are called the **reactants** and the things on the right are called the **products**. Sorta makes sense, because the things on the left *react* with each other to *produce* the things on the right.

Topic: Chemical Equations
Go to: *www.scilinks.org*
Code: CB020

(oxygen) + (methane) → (carbon dioxide) + (water)

Reactants **Products**

Next I'm going to replace the names of the chemicals with their molecular formulas, as in

$$O_2 + CH_4 \rightarrow CO_2 + H_2O$$

> **Coefficients and subscripts.** I've sort of explained this along the way, but it might help to summarize here. We use subscripts to show how many of each kind of atom are in a molecule. CO_2 has one carbon atom and two oxygen atoms. A number (the coefficient) in front of a molecular formula indicates how many molecules you have. $2CO_2$ is two molecules of carbon dioxide and $5CO_2$ is five molecules of carbon dioxide.

This doesn't accurately represent the reaction, though. If you'll notice, there are two oxygen atoms on the left and three oxygen atoms on the right. Also, there are four hydrogen atoms on the left and only two hydrogen atoms on the right. Did two of the hydrogen atoms combine and form an oxygen atom? Nope. Chemists discovered a long time ago that, in normal chemical reactions, matter is neither created nor destroyed. Hydrogen atoms have a lot less mass than oxygen atoms (just look at their molecular masses in the Periodic Table), so in order for two hydrogens to turn into one oxygen, we would have had to

Topic: Conservation of Mass
Go to: *www.scilinks.org*
Code: CB021

generate matter in the process. The fact that matter is neither created nor destroyed in chemical reactions is known as the **Law of conservation of mass.**[1] Mass is a measure of how much matter you have, which is why the word mass is in that law.

In order to alter our reaction so that we aren't creating or destroying matter, we have to balance the reaction, or develop a **balanced equation**. In order to do that, we just have to make sure that there are equal numbers of like atoms on either side. Balancing an equation is a puzzle that is often easy to solve, but sometimes difficult. I chose an easy one to start with. We have

$$O_2 + CH_4 \rightarrow CO_2 + H_2O$$

Let's just look at the hydrogen atoms. If we had two water molecules on the right, that would make four hydrogens on the right and four hydrogens on the left, which would balance out the hydrogen atoms, so let's do that.

$$O_2 + CH_4 \rightarrow CO_2 + 2H_2O$$

Notice that all I did was put a 2 out in front of the water, indicating two water molecules. Because each water molecule has two hydrogen atoms, we now have four hydrogen atoms on the left and four on the right, so that's good. Our oxygen atoms are still out of whack, though. There are two oxygen atoms on the left and four on the right (two in the carbon dioxide and two in the two water molecules). Easily fixed. All we have to do is multiply the oxygen molecule on the left by 2, as in

$$2O_2 + CH_4 \rightarrow CO_2 + 2H_2O$$

Now we have four oxygens on the left and four on the right, and the same number of each kind of atom on each side. Our equation is balanced. The fact that the number of each kind of atom on each side is the same is why we can refer to this as an equation. For completeness, I should write this equation as you often will see it, which is with parentheses to the right of each molecule indicating whether it's in a gaseous state (we use the letter *g*), dissolved in water (we use *aq* for "aqueous"), or a solid state (we use the letter *s*).

---

[1] This is a tiny lie. A very small amount of matter actually changes into energy in the form of heat during this reaction, which is why that methane burning in your gas stove gives off heat. This transformation of matter into energy is described by Einstein's theory of relativity, and we're not going to get into that here. For chemical reactions, the change in mass is so tiny that we can safely ignore it.

$$2O_2\ (g) + CH_4\ (g) \rightarrow CO_2\ (g) + 2H_2O\ (g)$$

The reason there's a *g* by the water is that this normally shows up as water vapor. That water vapor might very well condense into liquid form later. I should also mention that when you burn methane gas with your stove, you are involving many more than two oxygen molecules and one methane molecule. This chemical equation is therefore *representative* of what happens on a large scale.

As a double check on whether or not you have balanced an equation, you can make a list of the number of each kind of atom on the left and compare with the number of each kind of atom on the right. If the numbers come out the same, you're good to go. Here's that check for the equation we just balanced.

Left	Right
oxygen 4	oxygen 4
carbon 1	carbon 1
hydrogen 4	hydrogen 4

> Take a moment. Before we go on, it's worthwhile to pause and reflect on what we're doing. We are using the concept of atoms and molecules to describe a process in which one invisible substance (methane) combines with another invisible substance (oxygen) to form a third invisible substance (carbon dioxide) and a fourth invisible substance (water vapor). We know how many of each kind of molecule is involved. That's pretty remarkable, not so much that we can describe invisible things, but that this kind of representation has far-reaching power for understanding our world—even beyond the ideas of earth, air, fire, and water.

Let's move on to the candle you burned. When a candle burns, oxygen in the air combines with the paraffin in the candle. This produces carbon dioxide and water, just as with the burning of methane. The molecular formula for paraffin is $C_{28}H_{58}$, meaning that paraffin has 28 carbon atoms and 58 hydrogen atoms (big molecule!). We can represent the burning of paraffin as follows:

(oxygen) + (paraffin) → (carbon dioxide) + (water)

Substituting the molecular formulas gives

$$O_2 + C_{28}H_{58} \rightarrow CO_2 + H_2O$$

Clearly we don't have a balanced equation, so we have to start messing around with the numbers of molecules. For starters, let's try balancing the number of carbon and hydrogen atoms. There are 28 carbon atoms on the left and 58 hydrogen atoms on the left, so we'll just multiply our carbon dioxide molecules on the right by 28 and our water molecules on the right by 29 (there are two hydrogens in a water molecule).

$$O_2 + C_{28}H_{58} \rightarrow 28CO_2 + 29H_2O$$

This balances the carbon and hydrogen atoms, so let's look at the oxygen atoms. On the right, we currently have 56 oxygen atoms from the carbon dioxide molecules and 29 oxygen atoms from the water molecules, for a total of 85 oxygen atoms. At a glance we can tell that this won't work, because 85 is an odd number. The only oxygen atoms on the left are in the oxygen molecules, and with two atoms per molecule, any number of oxygen atoms on the left will always be an even number. So ... let's try multiplying our 29 water molecules on the right by two, giving us 58 total, as in

$$O_2 + C_{28}H_{58} \rightarrow 28CO_2 + 58H_2O$$

That's going to give us an even number of oxygen atoms on the right, so that might work. Before simply balancing out the oxygen molecules at this point, though, notice that we now have 116 hydrogen atoms on the right (58 water molecules that contain two hydrogen atoms each) and only 58 hydrogen atoms on the left. To balance the hydrogen atoms, we can multiply the number of paraffin molecules on the left by two.

$$O_2 + 2C_{28}H_5 \rightarrow 28CO_2 + 58H_2O$$

Okay. Now the hydrogen atoms are balanced, but the carbon atoms aren't. There are 56 carbons on the left and only 28 on the right. We can fix that by multiplying the carbon dioxide molecules on the right by 2 (2 times 28 is 56).

$$O_2 + 2C_{28}H_{58} \rightarrow 56CO_2 + 58H_2O$$

We're almost there. 56 carbon atoms on the left and 56 carbon atoms on the right. 116 hydrogen atoms on the left and 116 hydrogen atoms on the right. The only problem is the oxygen atoms. There are 2 oxygens on the left and 170 oxygens on the right (112 from the carbon dioxide and 58 from the water). Because 2 times 85 equals 170, we just have to multiply the oxygen molecule on the left by 85, ending up with

$$85O_2 + 2C_{28}H_{58} \rightarrow 56CO_2 + 58H_2O$$

The equation is now balanced. Why does it have to be balanced? Because you don't create or destroy atoms in a chemical reaction. All you do is rearrange the atoms to form different molecules.

I should point out that there are lots of oxygen, paraffin, carbon dioxide, and water molecules hanging around the apparatus for our activity. Not all of them get involved in the chemical reaction. For example, only a small portion of the paraffin in the candle burns, and not all of the oxygen molecules inside the glass are involved in that burning. The chemical equation we have above only applies to the molecules that actually get involved in the reaction and, once again, this reaction is only *representative* of the molecules that *do* interact. Clearly millions of molecules are involved, not just 85 oxygen molecules and 2 paraffin molecules.

I told you that the balanced equation could give us information about why the water rises up inside the glass after the candle flame goes out, and it does. For every 85 oxygen molecules removed from the air, you get 56 carbon dioxide molecules back in the air, and 58 water molecules. Now, if all those water molecules end up as water vapor (a gas), then that means we get an *increase* in gas molecules inside the glass (for every 85 oxygen molecules removed you get 56 carbon dioxide molecules *plus* 58 water vapor molecules produced). That's not what happens, though. It turns out that, as soon as you get a short distance from the flame, a substantial number of the water molecules produced end up as liquid water rather than water vapor (check for water droplets on the inside of the glass). So, there is a net *reduction* in the number of gas molecules inside the glass. This reduces the **air pressure** inside the glass, allowing the outside air to push the water up into the glass.[2] Earth, air, water, and fire couldn't have predicted that result!

Why balance equations? I just gave you one example of how it's useful to balance chemical equations, but maybe I can paint a bigger picture. When you make s'mores, you know that you need, say, four squares of chocolate, two marshmallows, and two graham crackers per s'more. If you want to make 20 s'mores, then you know exactly how many marshmallows, chocolate squares, and graham crackers you need. If you are, instead of making s'mores, making a certain kind of plastic, then it's useful to know how much of each kind of molecule will create how much of your product. So, you are answering the question, "How much of this and how much of that will produce what I want?"

[2] No, the water isn't "sucked" into the glass, but rather pushed by a difference in pressure. See the *Stop Faking It!* book on air, water, and weather for a thorough discussion of air pressure differences and how they apply to this particular activity.

More things to do before you read more science stuff

For this section, you can either repeat activities from Chapter 1 or just remember what happened there. I'm talking about mixing baking soda and vinegar and heating Epsom salts. To refresh your memory, one reaction fizzes a lot, and the other sputters, resulting in a weight loss. If you don't know which is which, you really ought to review Chapter 1!

More science stuff

I explained these two reactions using the four elements of Empedocles, and if you recall, using early models of the atom to explain the results paled in comparison to what we could explain using the four elements of earth, water, air, and fire. Our new improved model of chemical reactions will prove to be much more useful.

Baking soda, or *sodium bicarbonate,* has a molecular formula of $NaHCO_3$. Each molecule of baking soda has one sodium atom, one hydrogen atom, one carbon atom, and three oxygen atoms. Vinegar, also known as acetic acid, has a molecular formula of $HC_2H_3O_2$, so each molecule of vinegar has four hydrogen atoms, two carbon atoms, and two oxygen atoms.[3] When you bring these two together, a hydrogen atom from the vinegar jumps over to the bicarbonate molecule, and a sodium atom from the bicarbonate jumps over to replace the first hydrogen atom in the vinegar (yes, electronegativity values explain why this happens). You end up with a salt known as sodium acetate and another acid known as carbonic acid. Here's the chemical equation, first in words and then in symbols:

vinegar + sodium bicarbonate → sodium acetate + carbonic acid

$$HC_2H_3O_2 + NaHCO_3 \rightarrow NaC_2H_3O_2 + H_2CO_3$$

It turns out we don't have to do any balancing, as the numbers of atoms are equal on each side. The story's not done yet, though. The carbonic acid further undergoes a chemical reaction, separating into carbon dioxide and water.

carbonic acid → carbon dioxide + water

$$H_2CO_3 \rightarrow CO_2 + H_2O$$

[3] In case you're wondering why all the hydrogen atoms in the formula for vinegar aren't grouped together, as in H_4, that's because it's traditional to write the formula for acids in a special way that indicates the hydrogen atom they release easily. In fact, the easy release of hydrogen atoms is what makes such molecules acids, but you'll have to wait for the second chemistry book for a detailed discussion of acids and bases.

Again, we're balanced from the start. We've also explained the fizzing. We produce carbon dioxide gas, which bubbles up and escapes the liquid. This is much more detailed than saying that the original reactants contained a bit of earth, air, and water, but in a way it's quite similar. The water and carbon dioxide *were* sort of contained in the original reactants, and we just moved a few atoms around in order for them to appear. Maybe those Greek philosophers weren't so dumb after all.

On to the heating of Epsom salts, which turns out to be quite simple. Epsom salts as you get them from the store contain a salt, magnesium sulfate, with a bunch of water molecules attached. We write that as

$$MgSO_4 \cdot 7H_2O$$

The $MgSO_4$ is the magnesium sulfate, and the (·) indicates that there are seven water molecules attached to each magnesium sulfate.[4] When you heat Epsom salts, you gradually remove water molecules from the molecules. The sputtering you see is those water molecules becoming water vapor (the gaseous form of water), which is the same thing that happens when you heat a pan of water and the water gradually "disappears." And now we have explained the weight loss you observed. As you remove more and more water molecules from the Epsom salts, they weigh less. Simple, huh?

Even more things to do before you read even more science stuff

Get that track and marble you used in Chapter 3, and create the shape shown in Figure 5.3.

Figure 5.3

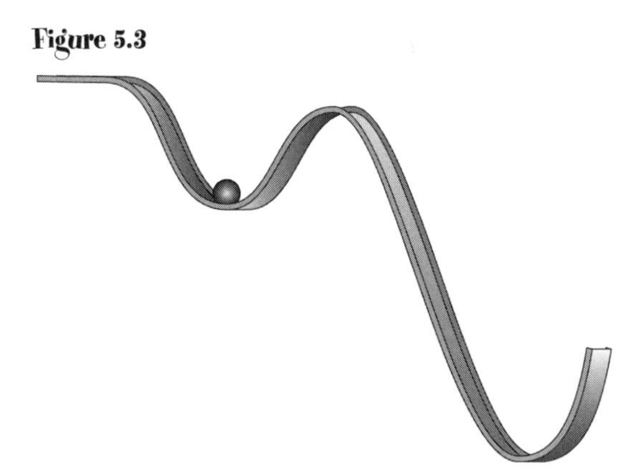

Place the marble in the upper trough. Move the marble until it falls into the lower trough. Now answer a simple question: What did you have to do in order to get the marble to gravitate to that trough of lower energy? Did you add energy to the marble or take it away?

[4] There are all sorts of rules for naming various combinations of atoms, leading to names like sulfide, sulfate, and carbonate. You can find those rules in any chemistry textbook, so I won't spend time discussing them here.

Even more science stuff

That was a pretty simple question. You had to *add* energy to the marble in order for it to be in position to head to the lower trough, which is at a lower overall energy level. The same sort of thing happens in chemical reactions. You had to add energy in the form of heat to remove the water from the Epsom salts. At minimum, you had to exert energy to mix the baking soda and vinegar so they would react. You have to add energy in the form of striking a match before the match burns, and you have to add the heat from the burning match in order for the paraffin in a candle to burn. Many other reactions require an input of heat to get the reaction "over the hump" so it can take place without any more input of energy. Many chemical reactions in cooking require this input of heat. If you just mix together the ingredients for a cake and don't put them in the oven, not much is going to happen.

Topic: Energy Conservation

Go to: *www.scilinks.org*

Code: CB022

Once you add a certain amount of energy, called the **activation energy**, to reactants, they are in position to get together and move to an overall lower energy. Remember, all physical systems tend toward the lowest energy possible. That's what's happening in chemical reactions, but you usually have to add energy first in order for the natural process to occur. This is probably a good thing, because it would be a strange world if all chemical reactions occurred spontaneously. You couldn't store baking soda and vinegar in the same pantry, because they would find a way to get together and produce sodium acetate, carbon dioxide, and water. Matches would light themselves. Not good.

Chapter Summary

- We can represent chemical reactions with "equations" that have molecular formulas for the reactants on the left and molecular formulas for products on the right.
- Balancing chemical equations is the process of making sure there are the same number of each kind of atom on the left and right. The law of conservation of mass tells us this relationship must hold true in any reaction.
- Balancing chemical equations allows us to determine how much of each reactant reacts with other reactants to produce a given amount of product. A balanced equation can provide insight into physical processes.
- Balanced chemical equations are only representative of the entire chemical reaction taking place.
- Some chemical reactions take place spontaneously, but most require an input

of energy, called the activation energy, for the reaction to proceed.

- A catalyst can facilitate a chemical reaction by lowering the activation energy for the reaction. Catalysts are not considered as final products of a reaction. They help a reaction proceed, but do not alter the final outcome other than the speed of the reaction.

Applications

1. Many chemical reactions require an input of energy in order to proceed. It's a matter of getting over the "energy hump" so the reaction can take place. You can do this by adding heat, for example. Another way is to add a different chemical, known as a **catalyst**, which speeds up the process of getting over the energy hump. Actually, the catalyst involves the reactants in a separate reaction that provides an alternative, and lower, energy hump to cross. Figure 5.4 shows an energy diagram for a reaction with and without a catalyst present. The nice thing about catalysts is that, although they make it easier for a reaction to occur, they themselves remain unchanged during the process and lead to the exact same products the reaction would have without the catalyst.

Topic: Catalysts
Go to: *www.scilinks.org*
Code: CB023

Figure 5.4

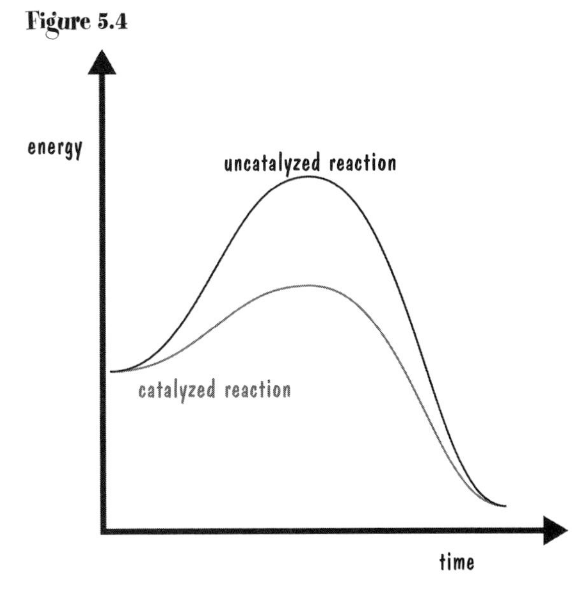

2. You undoubtedly know that photosynthesis is the process by which plants convert the Sun's energy to food in the form of sugars. Plants create their own food this way, and it's what makes plants a good thing to eat in order to survive. Something else happens during photosynthesis, and it's described by the following equation:

carbon dioxide + water → sugar + oxygen

$$CO_2 + H_2O \rightarrow C_6H_{12}O_6 + O_2$$

A quick glance at this equation will tell you it's not a balanced equation. We don't know how much of each molecule is involved in the reaction. The balanced equation turns out to be

$$6CO_2 + 6H_2O \rightarrow C_6H_{12}O_6 + 6O_2$$

What does this balanced equation tell us? That during photosynthesis, plants use carbon dioxide molecules and produce an equal number of oxygen molecules. This is what makes plants so useful to us other than as a food source. Because humans and other animals use oxygen and produce carbon dioxide, it's good to have plants that exchange carbon dioxide for an equal amount of oxygen. If the balanced equation showed that plants produced very little oxygen for a given amount of carbon dioxide, they wouldn't be nearly as useful in replenishing the supply of oxygen in the atmosphere, and human life might be difficult to sustain.

3. I left something out of the balanced equation for photosynthesis. I neglected to include the energy required to make the reaction happen. A better description of the reaction is:

$$6CO_2 + 6H_2O + \text{energy} \rightarrow C_6H_{12}O_6 + 6O_2$$

The amount of energy required to get this reaction to go forward is enormous. Where does that energy come from? The Sun—an essentially endless supply of energy. You go, Sun.

Organic, Dude

The word *organic* has lots of meanings in everyday life. Organic vegetables are ones grown without pesticides and organic beef comes from cattle that haven't been given antibiotics. When it comes to chemistry, *organic* sometimes means you are talking about living or once-living creatures. It also means something else, and you're about to find out what that is.

Things to do before you read the science stuff

Still have different-sized marshmallows around (Chapter 2)? Since there's only so many marshmallows you can roast over the stove, the answer is probably yes. And if those marshmallows are a bit stale, that's fine. Gather your marshmallows and a box of toothpicks. Break a number of the toothpicks in half. Stick a half-toothpick into one of the small marshmallows as shown in Figure 6.1.

Figure 6.1

This represents a hydrogen atom, and the half-toothpick indicates that, because hydrogen has a half-empty 1*s* shell, there is an opportunity for hydrogen to connect to another atom.

Build another hydrogen atom-marshmallow, complete with half-toothpick. Now pretend these two hydrogen atoms happen to meet at a party. We already know that hydrogen atoms can share electrons in a covalent bond, so simulate that by removing the half-toothpicks from the marshmallows and connecting the two hydrogens with a whole toothpick as shown in Figure 6.2. This represents the two hydrogen atoms sharing electrons so as to have complete energy levels—sort of.

Figure 6.2

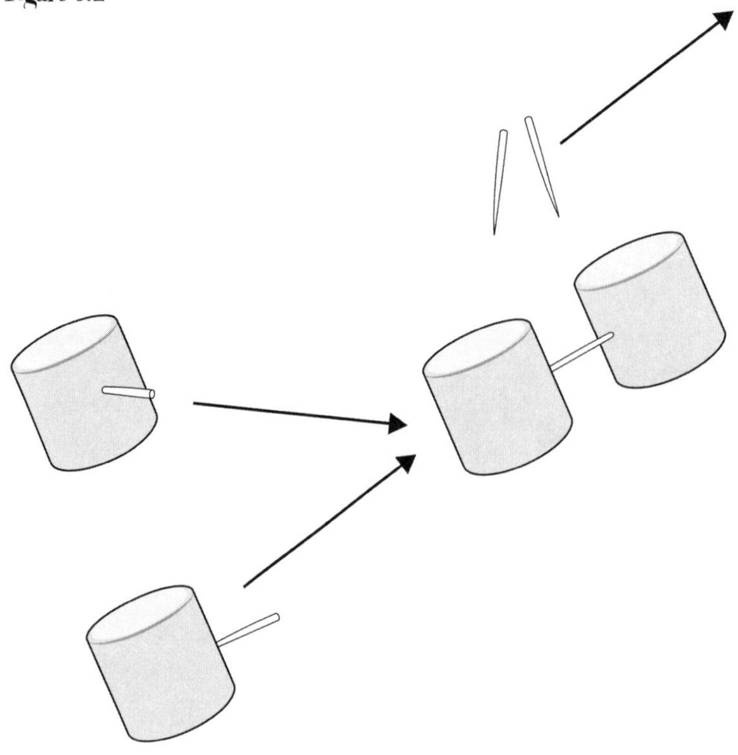

Now create a new marshmallow atom by placing two half-toothpicks in a large marshmallow, as shown in Figure 6.3. Once you've done that, create about four more of these atoms, along with another half dozen hydrogen atoms (small marshmallows with half-toothpicks extending from them.

Figure 6.3

The second kind of atom you have is supposed to be oxygen. This represents oxygen because there are two spaces (empty slots in an outer shell) where other atoms can connect.[1] Connect your hydrogen and oxygen atoms together in as many ways as you can. The bonds between these atoms are covalent, so you have to remove the half-toothpicks and replace with one whole toothpick when you make a connection to indicate the sharing of electrons. Remember that when two oxygen atoms get together, they often use both open slots in bonding together. Figure 6.4 shows a few possible connections.

Figure 6.4

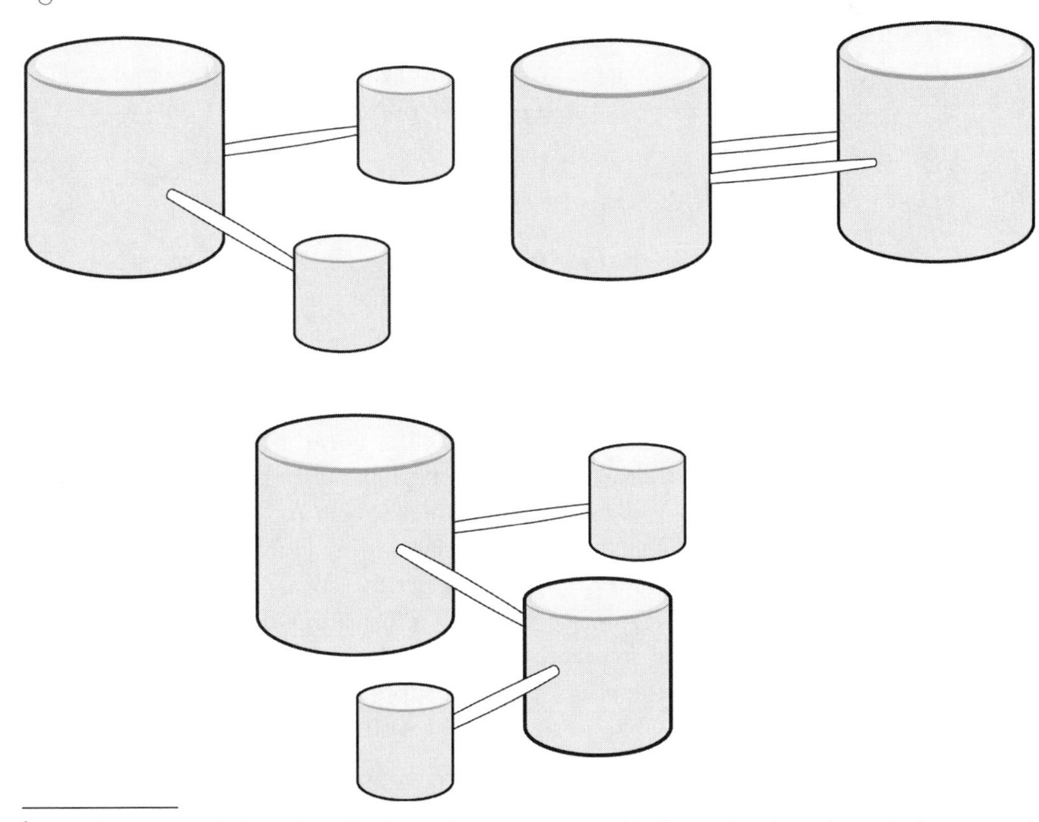

[1] You might be wondering why we've drawn the oxygen atom with the toothpicks at the angle they are. The reason is based on how the probability distributions for the electrons in oxygen interact. You'll just have to trust that this is how those probability distributions end up when oxygen bonds with other atoms. Yeah, it's quantum mechanics again!

How many different shapes of molecules can you create? Some, but not all that many. Take note of these different shapes so you can compare with what you create next.

Time to create a new atom—carbon. Take a large marshmallow and place four half-toothpicks in the marshmallow, as shown in Figure 6.5. This is a little tricky, because the probability distributions for electrons in carbon dictate that each distribution is equidistant from the distribution nearest it. Give it your best shot, because this accurately represents how carbon interacts with other atoms.

Figure 6.5

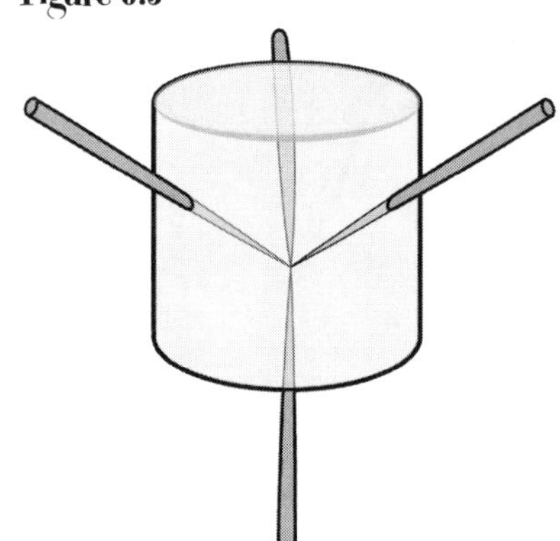

Create about a dozen of these four-toothpick (actually half-toothpick) atoms. Then use these, in combination with the marshmallow atoms you created earlier, to develop new molecules. Feel free to add more of the hydrogen atoms as needed. And *that's* why they have so many marshmallows in a bag!

How many different molecule shapes can you create now that you have the four-toothpick molecule? Limited only by your imagination, huh?

The science stuff

The marshmallow with the four toothpicks represents an atom of carbon. Carbon has four empty slots in its outer shell, and that makes carbon special. With four available slots to which other atoms can connect, carbon atoms can form all sorts of molecules, from simple ones such as methane to complicated ones such as 2,2,4-trimethylpentane, $(CH_3)_2CHCH_2C(CH_3)_3$, also known as isooctane, which is 100 octane gasoline.[2] Various possibilities of carbon bonding with other atoms are shown in Figure 6.6 (p. 91). The letters stand for different atoms, such as H for hydrogen, C for carbon, and O for oxygen. One dash means a single covalent bond (sharing one electron from each atom) and two dashes indicate a double covalent bond (sharing two electrons from each atom). Note that the drawings in Figure 6.6 are "flat" illustrations that don't reveal the three-dimensional structure of each molecule. Also, please realize that the molecule on the bottom right is supposed to be a joke—quit looking on the Periodic Table for an element with the symbol A.

SCiLINKS®
THE WORLD'S A CLICK AWAY
Topic: Basic Organic Chemistry
Go to: *www.scilinks.org*
Code: CB024

Molecules involving carbon are so important, the study of them has its own branch of chemistry known as **organic chemistry**. If you've ever taken a course in organic chemistry, you know that the chemical

[2] Just for fun, try balancing the burning of isooctane with oxygen to form CO_2 and H_2O.

Figure 6.6

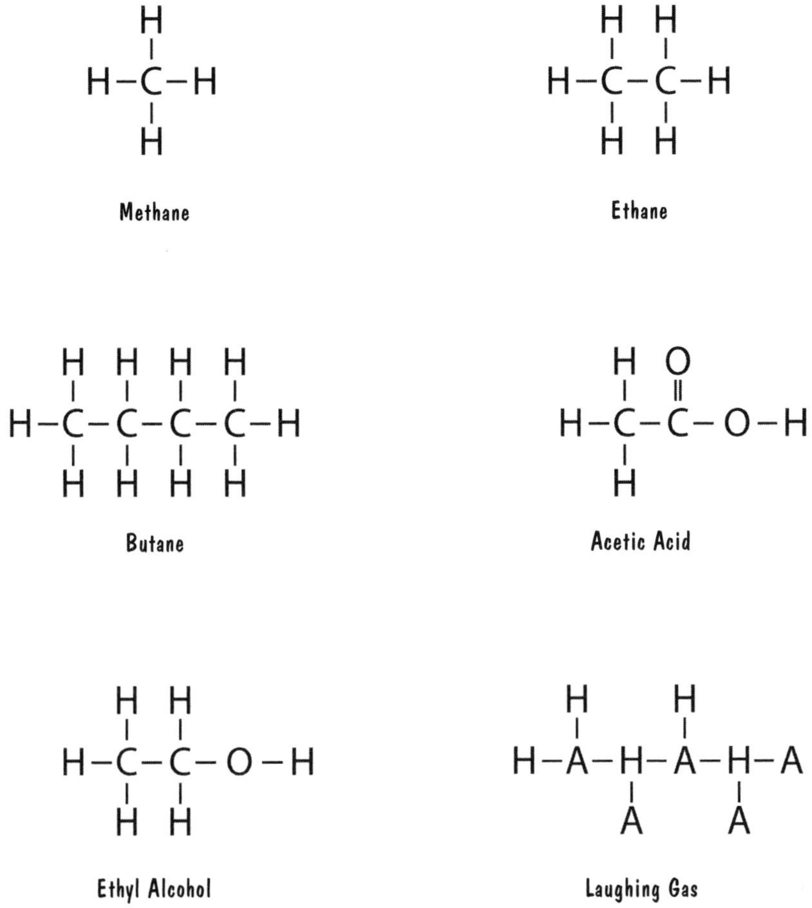

```
    H
    |
  H-C-H
    |
    H
```

Methane

```
    H H
    | |
  H-C-C-H
    | |
    H H
```

Ethane

```
    H H H H
    | | | |
  H-C-C-C-C-H
    | | | |
    H H H H
```

Butane

```
    H O
    | ||
  H-C-C-O-H
    |
    H
```

Acetic Acid

```
    H H
    | |
  H-C-C-O-H
    | |
    H H
```

Ethyl Alcohol

```
    H   H
    |   |
  H-A-H-A-H-A
      |   |
      A   A
```

Laughing Gas

reactions can get somewhat complicated, and it can seem like there are five million things to memorize, the most difficult of which can be the names of all the different molecules. Suffice to say we won't get into the subject that heavily here, and also suffice to say that there is a way of approaching organic chemistry so the memorization is at a minimum. Okay, okay, it's seldom taught with a minimum of memorization, but that's another story.

More things to do before you read more science stuff

Get an empty 2-liter bottle and head to the store and get a small box of borax, cleverly disguised as 20 Mule Team Borax, or other similar name. Also get a small bottle of white glue (not washable glue). Fill the 2-liter bottle with water and add about four tablespoons of borax. Shake this up and set it aside. In a separate container, mix about a quarter cup of school glue with an equal amount of water and stir. Notice the consistency of the glue solution and of the borax solution, because these are going to change dramatically.

Using an eye dropper or pipette, add small amounts at a time of boric acid solution to the glue solution, stirring as you go. Stop when you have something that resembles the material that comes in a plastic egg that you can buy at the toy store.

More science stuff

The molecules involved in what you just created are a wee bit complicated, but the reaction is instructive as to what kinds of things you can do in organic chemistry. First, though, some terminology. Many organic molecules are composed of repeating structures. The base structure is called a **monomer** and the chain of repeating structures is called a **polymer**. Makes sense, because polymer is a Greek word that means "many parts." The glue solution you used is a complicated mixture of molecules, but we can concentrate on the main ingredients. The first main ingredient is a polymer known as *polyvinyl alcohol*. It might be more instructive to write this as poly-(vinyl alcohol), since the base structure is the molecule vinyl alcohol, with a molecular formula of C_2H_4O. Polyvinyl alcohol can then be written as $[C_2H_4O]_n$, where the n can stand for any number greater than one, indicating that the base structure repeats many times.

Topic: Polymers
Go to: *www.scilinks.org*
Code: CB025

Attached to the polyvinyl alcohol in glue is another polymer known as polyvinyl acetate, which has a molecular formula of $[C_4H_6O_2]_n$. The vinyl acetate is the stuff that makes the glue a good adhesive once it dries. You didn't let it dry, though. Instead, you added another molecule to the picture—sodium tetraborate (this is the borax and water mixture you made). The formula for sodium tetraborate is $Na_2B_4O_7$, with two sodium atoms, four boron atoms, and seven oxygen atoms. The key parts of this molecule for the activity you did are the seven oxygen atoms. The way sodium tetraborate is put together, the oxygen atoms readily form a covalent bond with any hydrogen atoms that might be around. The two polymers, polyvinyl alcohol and polyvinyl acetate, are loaded with hydrogen atoms. What happens, then, is that the sodium tetraborate links together the polymers in the glue solution. This is called **cross-linking**. The glue solution, with its collection of polymers, is a lot like a box of separate metal chains that are free to move around. The sodium tetraborate links these chains together, so you end up with something like chain mail.[3] All the separate chains are glommed together, just like your final product when mixing glue and sodium tetraborate.

[3] I'm referring here to what medieval warriors used to wear, not one of those letters you have to answer and send out to other people or your cat will die.

Maybe right now you're asking, "So what?" Why care about making something rubbery out of glue? Well, the reason you might care is that a process similar to this is how we make all kinds of plastics and synthetic fabrics out of crude oil. It's just a matter of linking together the long polymers that make up oil in special ways.

Even more things to do before you read even more science stuff

I know that about now you're wondering when you're going to do an activity and then be able to eat the product of the activity. We aim to please. Following is a recipe for taffy.

Ingredients:

2 cups light corn syrup
1 cup sugar
2 tablespoons vinegar
2 teaspoons butter
Flavoring (vanilla, mint, cinnamon, or whatever you want)
Food coloring (optional)
A small bit of extra butter
1 cup of cold water

Liberally butter a baking pan. Mix together all of the ingredients except the food coloring and flavoring in a Teflon or similarly coated pot. Take a small amount of this mixture and put it in a second, smaller pot. Set this aside. Begin heating the original mixture until it comes to a boil. Do not stir the mixture while it's heating. Boil the mixture until a spoonful forms a ball when dropped in cold water (obviously, you have to keep testing it). If you have a candy thermometer, just boil until the mixture reaches a temperature of 250 degrees Fahrenheit. DO NOT HEAT THE MIXTURE BEYOND THIS TEMPERATURE. Pour the mixture into that buttered baking pan and add food coloring and flavoring if you want. Mix with a spatula and keep mixing until the mixture is cool enough to touch. Coat your hands with butter, take a handful of the mixture out of the pan, and begin pulling it into long strands, bunching it up, and pulling it into long strands again. Keep pulling on all the mixture for about five minutes (easier if you have helpers). Once it becomes almost impossible to pull, it's done. Form it into small pieces and wrap each one in waxed paper. Eat or sell on the street corner.

Still have that small bit of mixture in the other pot? Heat this up just as you did the large mixture, but heat it beyond the point where it forms a ball when dropped in cold water (heat it beyond 250 degrees). Remove from the heat and pour it into that buttered baking pan that's now empty. Let it cool, and then check whether or not you can pull it like you did the other stuff. Most likely not.

Even more science stuff

Making taffy is a great application of organic chemistry. The two main ingredients—corn syrup and sugar—are long-chain organic polymers. Corn syrup is composed of the molecule glucose, and table sugar is composed of the molecule sucrose. When you heat up these polymers, they, along with the butter,[4] break apart some of their old bonds and form new ones, creating a brand new polymer which is the taffy you eat. When you pull on the taffy, you stretch out these polymers into long strands that are easier to chew. And by the way, manufacturers go through a similar process in stretching polyesters into long threads that are used in fashion nightmares.

Okay, what about the mixture that you heated above 250 degrees? What you got out of that was a hard candy—still tasty but not the consistency of taffy. These high temperatures lead to longer, dehydrated (water molecules removed) polymers known as "char." These longer char polymers are crystalline in nature, which makes for hard candy.

Chapter Summary

- Carbon atoms play a special role in chemistry because their four bonding sites make it possible to form many kinds of molecules with carbon as the base atom. The study of carbon-based chemistry is known as organic chemistry.
- Because crude oil is carbon-based, many of the products we use everyday come from crude oil.
- A single base molecule of a long-chain molecule is called a monomer. The long-chain molecule is called a polymer.
- The long-chain nature of polymers, and cross-linking between polymers, enables many different products to be formed with polymers.

Applications

1. We use crude oil for lots of things, from plastics to synthetic fabrics to gasoline. When you go to the gas pump, you'll notice that different grades of gas have different *octane ratings*. Because octane is an organic molecule

[4] Butter is also composed of special organic molecules known as *esters*.

that burns quite well, you might expect that the octane rating on a grade of gasoline tells you how much actual octane is in the mixture. Well, you would be wrong if you expected that. The octane rating tells how efficiently the gasoline grade burns compared to the burning of octane. If the gasoline burns equally as efficiently as octane, it has an octane rating of 100. Obviously, then, since octane ratings of gasoline are from the mid '80s to the mid '90s, gasoline is not as efficient as pure octane.

2. Back in Chapter 1, I talked about the Greek philosopher Lucretius and his theory of how humans smell. It involved atoms of different shapes fitting into similarly shaped pores in the nose. Our current understanding of smell is not that different. First, most but not all molecules that we detect with our sense of smell are organic molecules. Organic molecules come in specific shapes, and the molecules in the nose also have specific shapes. One refers to these specially shaped molecules in the nose as *receptor sites.* When the molecules released from something like, say, vanilla beans reach the nose, they only fit into certain receptor sites, enabling us to identify different smells. Before they can get to these receptor sites, though, the molecules have to penetrate the mucous membranes contained in the nose. Because these membranes are mostly water, the molecules we smell have to be water soluble. If you refer back to Chapter 3, you'll notice that water soluble molecules are polar, meaning they have parts that are more positive or negative than the rest of the molecule. Of course, this explanation of the sense of smell goes out the window when dealing with noxious gases such as hydrogen sulfide (rotten egg gas). Even though they are not organic molecules, such gases are so highly reactive with other molecules that they have little trouble penetrating nasal membranes and messing with receptor sites in the nose.

SCILINKS®
THE WORLD'S A CLICK AWAY
Topic: Sensory Receptors
Go to: *www.scilinks.org*
Code: CB026

3. The fact that carbon so easily forms up to four covalent bonds with other molecules makes it perfect as the basis for a life form such as ours. Organic molecules readily mix and match and are able to transport energy needed for life. If you read or watch any science fiction, you have undoubtedly come across the notion that the universe can support silicon-based, rather than carbon-based, life. Why would the science fiction writers claim this? Just take a look at the Periodic Table and notice that silicon is right below carbon, meaning it has the same number of valence electrons and can form long-chain polymers, making it a good candidate for life's building blocks.

4. Many of the products we use are, in fact, silicon based. The caulking you use to seal up leaks is silicon based, as are those really bouncy rubber balls. There are other common uses of silicon, but I'll leave those to you to figure out. This is a family book.

What I haven't covered

There are a number of traditional chemistry topics not covered in this book, which is why there will be a second book on chemistry in the *Stop Faking It!* series. Here are a few topics you can look forward to seeing in the book that follows this one.

- Acids and bases and the pH scale
- Nuclear chemistry, including radioactive decay
- More detail on the interaction of light with matter
- More detail on orbitals and how they affect the formation of molecules
- Moles, molality, normality, and the like
- More on organic chemistry
- Chemistry in the human body
- Thermodynamics
- Electrochemistry
- Equilibrium
- Reduction-oxidation reactions
- Catalysts

Glossary

activation energy. The input of energy required to cause a chemical reaction to take place.

air. One of the four elements proposed by Empedocles. Also, a gas that is composed primarily of nitrogen and oxygen molecules. Also, the stuff we breathe.

alpha particle. A helium nucleus that has been stripped of its electrons and therefore contains two protons and two neutrons. As such, alpha particles are positively charged. Also, the particle that all other particles look to for leadership (think dogs).

atom. A tiny thing that has a positively charged nucleus surrounded by negatively charged electrons. Elements are made up of only one kind of atom. Also, a tiny little guy who had his own comic book for a while.

atomic mass. A number, measured in atomic mass units, that when truncated gives the total number of protons and neutrons in the nucleus of a specific atom. This number is not an integer because of naturally occurring isotopes.

atomic number. A number that tells you the number of protons in a specific atom.

atomic weight. A term chemists used to use when they really meant atomic mass. You will still find this term in use today.

balanced equation. A chemical equation that has the same number of each type of atom on either side. Also, an equation that has its life together.

Bohr. A physicist who first proposed a model of the atom that included discrete energy levels for electrons.

catalyst. A chemical that helps a chemical reaction occur more easily or more rapidly, while not technically part of the reaction as a product or reactant.

chemical equation. An equation that isn't really an equation in the mathematical sense, that shows what reactants come together to produce certain products in a chemical reaction.

conductor. Something that conducts electricity well. Metals are good conductors. So are people who have the ability to get people where they need to go on a train.

covalent bond. A bond between atoms in which the atoms share one or more electrons.

cross-linking. The process in which a group of molecules connects a bunch of long-chain polymers. What you do to succeed in a game of Red Rover.

Democritus. A Greek philosopher who was one of the first to come up with a theory of atoms.

diffraction grating. A device that separates light out into its individual frequencies.

earth. One of the four elements proposed by Empedocles. When capitalized, it's the big round thing we live on.

electric force. A force between different charges.

electromagnetic wave. A wave that is composed of changing electric and magnetic fields. Radio waves, microwaves, light, and x-rays are all electromagnetic waves. Also, what electrons use for surfing.

electron. A tiny little particle that mathematically has no size at all, and is negatively charged. We usually find electrons attached to atoms.

electron shell. A somewhat diffuse location in an atom in which there is room for a specific number of electrons. Also, what electrons pull into when disturbed by goings on in current events.

electron shielding. The process by which outer electrons are shielded from the positive charge of the nucleus by electrons that are closer to the nucleus.

electronegativity. A number assigned to each kind of atom that gives an indication of how strongly the atom attracts electrons when bonding with other atoms. An electric current that gives off bad vibes.

element. A substance with specific properties that cannot be reduced to a more basic substance. Elements are composed of a single kind of atom. Also, a car designed for surfers.

Empedocles. A Greek philosopher who proposed that the universe is composed of four elements—earth, air, fire, and water. His theory does a remarkably good job of explaining a number of physical observations.

energy. A rather abstract concept that is important for understanding how electrons in atoms behave and for understanding various chemical reactions. Something you're supposed to get from those bad-tasting gel packets you get at fitness stores.

energy levels. A series of energy "slots" that are available to electrons in atoms, and which exist whether or not they contain electrons.

fire. One of the four elements proposed by Empedocles. Something Beavis likes to say.

frequency. How often something happens. With waves, it's a measure of how often the oscillations occur.

inert gas. the name given to elements at the far right column of the Periodic Table. These elements have complete outer electron shells, and therefore don't interact much with other elements. Hence, the label of inert.

inner shell. An electron shell that is relatively close to the nucleus compared with the electron shells that are farther away from the nucleus. Of course, there actually must be outer shells in the atom for ones close to the nucleus to be called inner shells.

insulator. A substance that does not conduct electricity very well.

ion. An atom that contains extra electrons or has lost one or more electrons, and is thus charged rather than neutral.

ionic bond. A bond between different atoms in which an electron moves from one atom to the other, resulting in an electric force between the atoms.

isotope. An atom that has one or more "extra" (more than the usual number) neutrons in its nucleus. The extra neutrons do not change the identity of the atom.

Law of conservation of mass. The law that states that no mass is gained or lost in a chemical reaction. This is technically not true, but the gain or loss of mass in chemical reactions is so small that it is usually ignored.

Lewis dot formula. A visual method of showing how many valence electrons are in an atom and how those valence electrons end up when bonds form.

mass. A measure of the inertia of something, or the amount of matter contained in something.

mass spectrometer. A device that can be used to measure the charge to mass ratio of charged particles.

metal. A classification of elements that conduct heat and electricity well, and also have other properties in common. Most of the elements in the Periodic Table are metals. Also, a type of music that takes getting used to for us old folks.

Millikan. A physicist whose experiment successfully determined the charge on an electron.

molecular formula. A combination of element symbols and subscripts that tells how many of each kind of atom contribute to a given molecule.

molecular mass. The mass, in atomic mass units, of one molecule of a substance.

molecular weight. The term chemists used to use instead of molecular mass, though it is still commonly used today.

monomer. The basic unit of a polymer.

neutral atom. An atom with an equal number of positive and negative charges.

neutron. A particle with no charge that is found in the nuclei of atoms and has approximately the same mass as a proton. Also, Jimmy's last name.

noble gas. Another name for an inert gas. Also, the gas of kings and queens.

nonmetal. An element that does not have the properties of metals. These elements often have inferiority complexes because they are defined by what they are not.

nonpolar covalent bond. A covalent bond in which the distribution of shared electrons is such that the overall molecule does not have a preferred direction of charge.

nucleus. The concentrated center of an atom that consists of protons and usually neutrons. The base structure of the word *nuclear.* Note that there is no "u" between the c and the l in *nuclear.*

orbital. The name given to the probability distribution associated with various electron energies in atoms.

organic chemistry. The branch of chemistry that deals with the myriad of molecules that contain carbon. Also, chemistry done without pesticides.

oscillation. Any periodic variation in time.

outer shell. An electron shell that is farthest away from the nucleus in an atom.

Pauli exclusion principle. A principle that states that no two electrons in an atom can have exactly the same set of quantum numbers.

Periodic Table. A large table that contains all of the elements and illustrates a number of patterns relating to the structure of atoms and the properties of elements. Also, the card table that comes out on holiday dinners when you need an extra place for the kids to sit.

photosynthesis. The chemical process by which plants convert solar energy to food.

polar covalent bond. A covalent bond in which the distribution of shared electrons results in an asymmetric charge distribution.

polymer. A molecule that consists of repeated basic units. Many mers.

probability distribution. A visual representation of the probability of finding something in various places. An especially useful tool in determining what electrons in atoms are doing.

product. One of the molecules that results from a chemical reaction. Something you need if you're going to make money on eBay.

proton. A positively charged particle found in the nucleus of every atom.

quantum mechanics. A mathematically based theory in physics that does a great job of explaining how things behave on an atomic level. Also, a group of people who are trained in repairing quanta (bad joke I learned in college).

quantum number. A number that describes various energy levels in atoms.

reactant. One of the molecules that serve as the beginning point for a chemical reaction. Reactants go on the left in a chemical equation.

Rutherford. A physicist who conducted an experiment that showed there was a concentrated positive charge in atoms. Also, Lumpy's last name.

sea of electrons. An expression that describes how valence electrons behave in metals. They are shared by all the atoms in the metal and roam all over the place.

shell. The energy level in an atom.

standing waves. The pattern that results at certain frequencies when waves are confined. Also, a name for one of the characters rejected for the film *Dances With Wolves* when they realized waves weren't involved in the film.

Thomson. The scientist who came up with a model of the atom that had positive and negative charges distributed evenly throughout the atom.

valence electron. Any electron in the outer *s* or *p* energy levels.

water. One of the four elements proposed by Empedocles. Something you pay a lot for if it comes in a plastic bottle.

Index

Page numbers in **boldface** type refer to figures.

D

E

F

G

H

O

P

T

V

W

Z

Stop Faking It!

Don't struggle to teach concepts you don't fully understand... Instead, laugh while you learn with other award-winning books by NSTA Press best-selling author Bill Robertson.

Air, Water, & Weather

Air pressure, the Coriolis force, the Bernoulli Effect, density, why hot air doesn't rise by itself and why heating air doesn't necessarily cause it to expand.
PB169X6; ISBN: 978-0-87355-238-7; 134 pages

Electricity & Magnetism

The basics of static electricity, current electricity, and magnetism; develops a scientific model showing that electricity and magnetism are really the same phenomenon in different forms.
PB169X5; ISBN: 978-0-87355-236-3; 161 pages

Energy

Easy-to-grasp explanations of work, kinetic energy, potential energy, and the transformation of energy, and energy as it relates to simple machines, heat energy, temperature, and heat transfer.
PB169X2; ISBN: 978-0-87355-214-7; 106 pages

Force & Motion

Combines easy-to understand explanations and activities to lead you through Newton's laws to the physics of space travel.
PB169X1; ISBN: 978-0-87355-209-7; 100 pages

Light

Uses ray, wave, and particle models of light to explain the basics of reflection and refraction, optical instruments, polarization of light, and interference and diffraction.
PB169X3; ISBN: 978-0-87355-215-8; 115 pages

Math

This book takes a unique approach by focusing on the reasoning behind the rules, from math basics all the way up to a brief introduction to calculus: why you "carry" numbers, common denominators, cross-multiplying, and pi.
PB169X7; ISBN: 978-0-87355-240-0; 190 pages

Sound

Starting with the cause of sound and how it travels, learn how musical instruments work, how sound waves add and subtract, how the human ear works, and why you can sound like a Munchkin when you inhale helium.
PB169X4; ISBN: 978-0-87355-216-5; 107 pages

To read excerpts and to purchase these, and other resources from NSTA Press, visit ***http://store.nsta.org***

NSTA Members receive a discount!

Stop
Faking It!
SOUND